# GLOBAL WARMING AND AGRICULTURE

# GLOBAL WARMING AND AGRICULTURE

*Edited by:*

**Stephanya Lynn JonasLabee**

Delve
PUBLISHING

www.delvepublishing.com

**Global Warming and Agriculture**

*Stephanya Lynn JonasLabee*

**Delve Publishing**

2010 Winston Park Drive,

2nd Floor

Oakville, ON L6H 5R7

Canada

www.delvepublishing.com

Tel: 001-289-291-7705
 001-905-616-2116
Fax: 001-289-291-7601

Email: orders@arclereducation.com



ISBN: 978-1-77407-242-4 (Hardcover)

# ABOUT THE EDITOR

**Stephanya Jonas-Labee** has been passionate about the environment her whole life. After completing her bachelor's at UCSC, she moved to Amsterdam to earn her Master's of Earth Science with a focus on environmental management and sustainability from the University of Amsterdam. When not reading about new sustainable technologies, she enjoys traveling the world, trying out new vegan recipes, and working on her garden.

# TABLE OF CONTENTS

*List of Figures* ........ *xi*

*List of Tables* ........ *xiii*

*Preface* ........ *xv*

**Chapter 1 Introduction to Global Warming ........ 1**

1.1 Introduction ........ 2

1.2 The Relationship Between Global Warming and Climate Change ........ 2

1.3 Global Warming Causes ........ 4

1.4 Relationship Between Ozone Layer Depletion and Global Warming .. 5

1.5 Global Energy Demand and Impacts of Global Warming ........ 10

1.6 Consequences of Global Warming ........ 15

1.7 Relationship Between Global Warming and Bushfire ........ 17

1.8 Effects of Sun Ionizing Radiation Due to Global Warming ........ 19

1.9 Urbanization and Global Warming ........ 23

1.10 Reevaluating the Coal Pollution ........ 24

1.11 The Burden of Global Warming ........ 27

1.12 Who Should Bear the Burden of Fighting Global Warming? ........ 28

**Chapter 2 Global Warming, Water, and Agricultural Crop Yields ........ 35**

2.1 Introduction ........ 36

2.2 Soil Water Availability and Crop Farming ........ 36

2.3 Global Warming Impacts on Rainfall Availability For Agricultural Practices ........ 38

2.4 Agro-Climatic Zones ........ 40

2.5 Surface Water Hydrology ........ 41

2.6 The Impact of Global Warming on Water Chemistry ........ 43

2.7 Plant Distribution and Water Relations ........ 45

2.8 Soil Water Availability ........ 47

2.9 Importance of Plant Water Relations To Production........ 47

2.10 Soil Moisture Stress on Photosynthesis ........ 52

**Chapter 3 Environmental Sustainability, Global Warming, and Agricultural Land Degradation........ 55**

3.1 Introduction........ 56

3.2 Three Fundamental Pillars on Social Development of Sustainable Development ........ 56

3.3 Actions to Ensure Environmental Sustainability ........ 58

3.4 Deforestation and Global Warming ........ 61

3.5 Resources Mismanagement and Global Warming........ 63

3.6 Impacts of Population Growth Towards Environmental Degradation . 65

3.7 Agricultural Practices and Environmental Degradation ........ 68

3.8 Global Warming and Agricultural Land Degradation ........ 69

3.9 Politics and Agricultural Land Degradation........ 70

3.10 The Way Forward Towards Natural Resource Management ........ 71

**Chapter 4 Relationship Between Global Warming and Agricultural Crop Farming ........ 75**

4.1 Introduction........ 76

4.2 Effect of Increased Temperature on Growth and Development of Agricultural Crops ........ 76

4.3 Impacts of Soil Salinity on Crops Due to Global Warming ........ 79

4.4 Impacts of Global Warming on Agricultural Soil Properties........ 82

4.5 The Connection Between Minimum Tillage and Global Warming ........ 87

4.6 Excessive Use of Agricultural Fertilizer Leads to Global Warming ........ 89

4.7 Transforming Agricultural Practices ........ 91

**Chapter 5 The Impacts of Global Warming on Key Agricultural Crops ........ 95**

5.1 Introduction........ 96

5.2 Maize Farming........ 96

5.3 Potato Farming........ 97

5.4 Tea Production........ 100

5.5 Sugarcane Farming ........ 102

5.6 Rice Production........ 103

5.7 Pigeon Pea Farming ........ 105

5.8 Pyrethrum Farming ..................................................... 114

**Chapter 6 Global Warming and Livestock Farming .......................... 117**

6.1 Introduction ..................................................... 118

6.2 The Impacts of Global Warming on Dairy Farming ..................... 118

6.3 The Effects of Global Warming and Climate Change on Beef Farming ..... 123

6.4 Goats Farming ..................................................... 131

6.5 Poultry Farming ..................................................... 134

6.6 Can Genetically Modified Animals Help In Alleviating The Effects of Global Warming? ..................................... 137

**Chapter 7 Relationship Between Impacts of Global Warming and Various Agricultural Pests and Diseases ..................................... 141**

7.1 Introduction ..................................................... 142

7.2 Livestock Ticks ..................................................... 142

7.3 Brucellosis In Livestock ..................................................... 147

7.4 Camel Trypanosomiasis ..................................................... 147

7.5 Salmonellosis ..................................................... 149

7.6 Cassava Mosaic Disease ..................................................... 149

7.7 Cereal Storage Insect Pests ..................................................... 151

7.8 Major Diseases And Pests of Coffee ..................................... 152

7.9 Cowpea Rust ..................................................... 155

7.10 Major Diseases And Pests of Beans ..................................... 157

7.11 Plant Parasitic Nematodes ..................................................... 159

7.12 Southern Leaf Blight ..................................................... 160

**Chapter 8 Economic Importance of Irrigation in Agricultural Sector With Reference To Global Warming ..................................... 163**

8.1 Introduction ..................................................... 164

8.2 Definition and The Need of Irrigation in Agricultural Sector ............ 164

8.3 Major Sources of Irrigation Water ..................................... 164

8.4 Benefits of Irrigation ..................................................... 165

8.5 Factors Influencing the Appraisal of Irrigation ..................... 167

8.6 Methods of Irrigation ..................................................... 170

**Chapter 9 Man's Response to the Impact of Global Warming ........ 175**
9.1 Introduction ........ 176
9.2 Energy Management ........ 178
9.3 Nuclear Power Energy Application ........ 181
9.4 Solar Energy as an Effective Renewable Energy ........ 185
9.5 Rain Water Harvesting ........ 189
9.6 Agricultural Crop Adaptation to Effects of Global Warming ........ 193
9.7 Insurance and Compensation: Agricultural Products ........ 195
9.8 Afforestation as a Mitigating Effort Towards Global Warming Impacts ........ 198
9.9 Reforming The International Institutions That Deal With Impacts of Global Warming and Climate Change ........ 201
9.10 Key Challenges in Mitigating the Impacts of Global Warming ........ 206

**Chapter 10 Global Environmental Policy and Management to Deal With Impacts of Global Warming ........ 209**
10.1 Introduction ........ 210
10.2 International Environmental Agreements ........ 210
10.3 Economy, Environment and The Social Life of Individuals ........ 213
10.4 Environment Impact Assessment (EIA) ........ 213
10.5 Steps In Environmental Impact Assessment ........ 213
10.6 Policy Evaluation ........ 214
10.7 Advantages of the Global Environment Management Agreements ........ 217
10.8 Challenges That Nations are Facing When Trying to Comply With The International Agreements ........ 218
10.9 The Roles of International Financial Organizations in Curbing Environmental Challenges ........ 219
10.10 Ethical Decision Making on Environmental Issues ........ 220

**References ........ 221**

**Index ........ 227**

# LIST OF FIGURES

**Figure 1.1:** Death of livestock due to the impact of global warming

**Figure 1.2:** Fossil fuel is the leading cause of global warming

**Figure 1.3:** Hydraulic fracturing

**Figure 1.4:** An illustration of how urbanization has contributed to global warming

**Figure 1.5:** Burning coal is the leading cause of global warming in the world

**Figure 2.1:** Global warming effects and climatic elements are the main sources of great potential and constraints to plant growth

**Figure 2.2:** Effects of soil moisture stress on plant photosynthesis

**Figure 2.3:** Structural adaptation in plants due to water shortage

**Figure 3.1:** Fundamental pillars on social development of sustainable development

**Figure 3.2:** Waste recycling and reusing

**Figure 3.3:** Current rate of deforestation is contributing a lot in terms of global warming.

**Figure 3.4:** High population growth rate has led to a high rate of deforestation

**Figure 4.1:** In tropical countries, agricultural plants are usually subjected to severe water stress at high light intensity due to the effect of global warming

**Figure 4.2:** Some of the major impacts of soil salinity on crop plants may range from impaired germination to the death of the plants

**Figure 4.3:** Minimum tillage

**Figure 4.4:** Excessive use of agricultural fertilizer leads to global warming

**Figure 5.1:** Potato production has declined due to the impact of global warming

**Figure 5.2:** High maximum temperatures in excess of 29.5 °C have damaged most of the tea plantations in most areas

**Figure 5.3:** It seems that the frequent moisture stress experienced in the Sugar Belt may largely account for the poor and stunted growth in sugarcane

**Figure 5.4**: Drought conditions threaten rice farming in most places around the

world
**Figure 5.5:** Pigeon peas that dried up before they could be harvested
**Figure 5.6:** New pigeonpea cultivars have been developed to curb the effect of global warming
**Figure 6.1:** The impacts of global warming affect dairy farming differently
**Figure 6.2:** Traditional pastoralism beef farming
**Figure 6.3:** Commercial ranch
**Figure 6.4:** Emaciated beef cows due to the effects of global warming
**Figure 6.5**: Beef cows are more exposed to the total impact of global warming
**Figure 6.6:** Goat farming is also experiencing difficulties due to impacts of global warming
**Figure 6.7:** Despite severe climatic conditions, goats are able to survive because they can browse as well as graze
**Figure 6.8:** The DNA and RNA structures that show that the interconnection of the four bases
**Figure 7.1:** Ticks infestation on livestock
**Figure 7.2:** Camel are severely affected by trypanosomiasis due to effects of global warming in desert areas
**Figure 7.3:** Whiteflies
**Figure 7.4:** Major pests of stored cereals
**Figure 7.5:** Coffee berry disease
**Figure 7.6:** Coffee leaf rust
**Figure 7.7:** Coffee mealbug infestation
**Figure 7.8:** Cowpeas rust as a result of adverse effects of global warming
**Figure 7.9:** Adverse impact of global warming on cowpeas production
**Figure 8.1:** A typical irrigation system in arid and semi-arid areas
**Figure 9.1:** The global consumption of energy in the world has been increasing year after year
**Figure 9.2:** The trend in nuclear power production by country
**Figure 9.3:** Solar power application
**Figure 9.4:** A typical rainwater harvesting system
**Figure 9.5:** In situ rainwater harvesting technique
**Figure 9.6:** Some of the major agricultural products to insure

# LIST OF TABLES

**Table 1.1:** Effects of ozone depletion on human beings

**Table 1.2:** Features of global storm

**Table 2.1:** Effects of the water balance on growth and development of plants

**Table 7.1:** Seasonal incidences of some species of ticks

**Table 8.1:** Variable factors which influence costs of irrigation

**Table 9.1:** Utilization of solar energy in food processing industry

**Table 10.1:** Instruments for managing and protecting the environment

# PREFACE

Numerous research studies related to the impacts of global warming and agriculture have been done in different parts of the world. There exists a closer association between global warming and agriculture. However, there is a need for more information on these two aspects in order to help in mitigating the causes of global warming and ensuring that there is food security in the world. At the global level, different policies and other international declarations have been created to help alleviate the impacts of global warming in the world. Debates have been made to consider the ramifications of environmental change on an assortment of agricultural practices including management of natural resources such as water, forests, and agricultural lands.

Information on the impacts of global warming and agricultural practices is very variable. Several studies related to these two aspects have been carried out and presented in different books and reports. However, the information is dispersed and does not offer a comprehensive overview of the real problem of global warming and agricultural practices. Therefore, this book was written after doing thorough scrutiny of different journals, articles, research, and a wide array of global warming and agricultural practices books. The main objective of preparing this book is to offer the best information based on the field of global warming and agriculture.

Different readers can make use of this book including undergraduate and postgraduate students, scientists , research institute employees, planners, implementing agencies, and also various governmental and non-governmental agencies dealing with the environmental and agricultural sector.

The book has ten chapters, and each chapter discusses a specific topic under global warming and agriculture: chapter one discusses about introduction to global warming; chapter two is about global warming, water, and agricultural crop yields; chapter three discusses about environmental sustainability, global warming, and agricultural land degradation; chapter four discusses the relationship between global warming and agricultural crop farming; chapter five presents information about the impacts of global warming on key agricultural crops; chapter six discusses global warming and livestock farming; chapter seven offers detailed information about the relationship between the impacts of global warming and various agricultural pests and diseases; chapter eight offer information on economic importance of irrigation in the agricultural sector with reference to global warming; chapter nine discusses man's response to the impact of global warming; and finally, chapter ten is all about comprehensive information on global environmental policy and management to deal with the impacts of global warming.

These ten chapters were lengthily prepared for the readers to have a preview of how global warming and agriculture relate to each. The author paid more emphasis on the simplicity of the content in order to make the book comprehensible and understandable to all readers.

*Editor*

# Introduction to Global Warming

# 1

CONTENTS


1.1 Introduction .......... 2
1.2 The Relationship Between Global Warming and Climate Change .......... 2
1.3 Global Warming Causes .......... 4
1.4 Relationship Between Ozone Layer Depletion and Global Warming .......... 5
1.5 Global Energy Demand and Impacts of Global Warming .......... 10
1.6 Consequences of Global Warming .......... 15
1.7 Relationship Between Global Warming and Bushfire .......... 17
1.8 Effects of Sun Ionizing Radiation Due to Global Warming .......... 19
1.9 Urbanization and Global Warming .......... 23
1.10 Reevaluating the Coal Pollution .......... 24
1.11 The Burden of Global Warming .......... 27
1.12 Who Should Bear the Burden of Fighting Global Warming? .......... 28

## 1.1 INTRODUCTION

Global warming, also called climate change, refers to temperature rises on Earth. On an average, the global temperatures have increased by 1.4° F in the last 100 years, with some quite accelerated temperature increases since 1975. The average surface temperature of the earth has increased by close to 1.62 degrees Fahrenheit since the late 19th century. Since 2010, records show that the plat was warmest over 5 years (NASA: https://climate.nasa.gov/evidence/).

As a result of global warming, there has been a troublesome melting of glaciers, the rise of sea levels, diminishing of forests, floods, droughts, and prolonged dry spells. Unless the global community agrees on how to stop further climate change, global warming will cause disastrous consequences on human social life, the economy, and even pose a threat to human existence.

This heat (0.4 degrees Fahrenheit) has been absorbed by above 700 meters of the sea (Levitus et al, 2017). Close to 127 billion tons of ice has been lost from Antarctica between 1993 and 2016 while the last century has seen an increase by 8 inches in the sea level this century (NASA: https://climate.nasa.gov/evidence/).

## 1.2 THE RELATIONSHIP BETWEEN GLOBAL WARMING AND CLIMATE CHANGE

Global warming is the projected slow increase in the general temperature of the Earth's atmosphere and oceans. On the other hand, climate change is a long-term change in the global average climate. Despite the fact that these two are connected in the aspect that global warming causes climate change, global warming differs from climate change. This is because global warming is a worldwide phenomenon while climate change can be observed at 3 levels: worldwide, regional level and even local .

Global warming hastens climate change. The effects of global warming have the potential to cause dramatic and long-term climate change. Most, if not all, of these global warming effects, are due to the contribution of actions of mankind which eventually lead to changes in the climate. Climatic changes are disruptive. A lot of problems have been experienced as a result of global warming and climate change. The gradual overall rise in temperature results in rising ocean temperatures and sea levels which easily leads to floods and severe coastal weather. These conditions put the coastal populace at high risk of displacement or even worse, the death of people and other land-based living organisms. The 2017 Atlantic hurricane, according to Scientific American, was brought about by climate change which ended up being a record-breaking hurricane. The Gulf Coast, U.S. eastern seaboard and most Caribbean islands suffered the most (Anderson, 2015).

**Figure 1.1:** Death of livestock due to the impact of global warming.

**Source:** https://www.smithsonianmag.com/science-nature/eight-ways-climate-change-hurts-humans-180950475/

The primary source of energy which drives the climate system of the earth is the sun. However, recent studies suggest that there is a minimal effect caused variation in the sun's energy. In an attempt to find the most probable cause of climate change, the Inter-governmental Panel on Climate Change (IPCC) was formed by the United Nations. According to the fifth IPCC report, humans are the major contributors to the current experienced global warming, hence the climate change. The greenhouse gases emitted through various human activities are majorly responsible for warming in about 95% over the last 60 years. There has been a constant increase in the emission of fossil fuel since the industrial revolution (IPCC, 2014).

The greenhouse gases have been estimated by IPCC to have caused a global mean surface warming of between 0.5-1.3 °C from studies conducted between the year 1951 and 2010. Other anthropogenic factors which include the cooling effect of aerosols have contributed between -0.6 to 0.1 °C. About 50-150% of the global warming observed from 1950-2005 is caused by human climate influences. These gases include carbon dioxide (which is the main component), methane, and nitrous oxide among others. Human activities have disturbed the natural carbon cycle through the release of carbon dioxide to the atmosphere. Human activities have interfered with the carbon cycle. Through releasing of greenhouse gases such as chlorofluorocarbons, a blanket of these gases has formed in the atmosphere leading to an increase in global temperatures. This, in turn, leads to climate change that is being observed today (Anderson, 2015).

## 1.3 GLOBAL WARMING CAUSES

Modern human activities are the main cause of global warming that include:

- Mining
- Industrial production
- The burning of fossil fuels
- Deforestation
- Urbanization

Mining has negative effects on the climate. During the mining process, methane becomes entrapped in the Earth's belly, only to cause heat waves later. Even the rearing of cattle can enhance global warming because livestock manure comprises methane, a gas that causes global warming. Human industrial activities contribute to climate change too. Most machines in factories run on fossil fuel. For instance, garment, car, and mobile making industrial plants release a lot of carbon dioxide. Coupled with the automobile fuel exhausts, the carbon dioxide can bring heat waves and extremes temperature fluctuations.

The discharge of greenhouse gases in the last century means the existence of "heat-trapping gases" in the atmosphere. Consequently, the world is warmer than it was 650,000 years ago. Moreover, the greenhouse gases have caused a change in weather patterns, which is the reason the world experiences the fury of *El Niño* (prolonged and heavy rainy season) or *La Nina* (long sunny spell). People often cut down trees for house construction, paper manufacture, fuel, and furniture. As Weaver (2011) notes, trees play the crucial role in absorbing carbon dioxide from the atmosphere. However, the absence of an adequate forest cover to absorb emitted gases means that carbon dioxide will remain in the atmosphere and damage the ozone.

Eventually, the heat of the sun will burn the bare earth surface more, cause wanton water evaporation, and droughts. Global warming can be amplified by urbanization and the built environment. Rapid urbanization will enhance "increases thermal-storage capacity, known as the Urban Heat Island (UHI) effect." Road infrastructure and cities have a significant contribution to UHI and climate change.

## 1.4 RELATIONSHIP BETWEEN OZONE LAYER DEPLETION AND GLOBAL WARMING

Depletion of the ozone layer is one of the serious challenges planet earth is facing currently. The ozone layer is made up of a colorless gas referred to as the ozone gas ($O_3$), found in the stratosphere of the upper atmosphere. 93-99% of the sun's high frequency ultraviolet light is absorbed by this layer that contains 91% ozone located in the 10 km to 50 km above Earth: the stratosphere (Sivasakthivel and Reddy, 2011).

This gas is significant because it protects us from the harmful effects of the ultraviolet radiations of the sun. This layer protects all the living things on the other by absorbing the harmful radiations preventing them from entering the earth's atmosphere. The ultraviolet radiations are electromagnetic waves that are highly charged emitted by the sun. When these waves enter the atmosphere of the earth, they result in dangerous environmental challenges which include global warming and health issues related to the living organisms on earth (Anderson, 2015). The ozone layer is thus a significant aspect of protecting people from the harmful sun rays.

### 1.4.1 Major causes of Ozone Layer Depletion

*Chlorofluorocarbons (CFCs):* The depletion of the ozone layer is caused by various human factors. First, chlorofluorocarbons (CFCs) are compounds used by man in his daily life. These compounds are dangerous to the ozone layer thus leading to its depletion. The compounds contain chlorine, fluorine, and carbon elements. The compounds are mainly used as solvents and refrigerants (Anderson, 2015). The most common compounds include:

- CFC-11
- CFC-12
- CFC-113
- CFC-114

- CFC-115

All these 5 have depletion values of 1, 1, 0.8, 1, and 0.6 respectively. These compounds react with ultraviolet radiation when they reach the stratosphere of the earth. The ultraviolet radiation disintegrates the compounds causing them to release bromine and chlorine gases into the ozone layer of the earth. These harmful chemicals released to cause a series of reactions in the ozone layer disintegrating it to become pure oxygen gas. As a result, the ozone layer loses its protective effect from the UV rays emitted by the sun. The chlorine and bromine molecules continue reacting with the ozone molecules putting the ozone layer at the risk of bonding with other molecules (Anderson, 2015).

*Hydrochlorofluorocarbons (HCFCs):* Hydrochlorofluorocarbons (HCFCs) are also harmful compounds to the ozone layer. These compounds have been lately used to substitute CFCs because they have similar properties. Another reason for the substitution is because HCFCs are less harmful to the ozone layer because they have a short half-life and they release fewer chlorine atoms when compared to CFCs (Anderson, 2015). The decrease in the ozone layer depletion when HCFCs are used ranges between 0.01 and 0.1. Despite the fact that they are better than CFCs they remain harmful to the ozone layer. The use of HCFCs is considered to be a temporary solution with the developing countries banning their use from 1930.

*Halons:* Like HCFCs and CFCs, halons are compounds used by man in daily life which affects the functioning of the ozone layer. Halons are compounds formed by reactions involving bromine, fluorine, and carbon. These compounds have the ability to extinguish a fire and thus they have been widely used in the manufacture of fire extinguishers (Casper, 2010). Despite their fire extinguishing ability, many countries in the world have prohibited their manufacture because they deplete the ozone layer. These compounds are extremely dangerous to the ozone layer due to the presence of bromine gas in their composition. Bromine atoms are more effective in destroying the ozone layer at a higher rate than chlorine atoms. In a nutshell, halon 1301 and halon 1211 depletion values on the ozone layer of 13 and 4 respectively.

*Pesticides and fertilizers:* Farming activities have also contributed to an increase in the rate at which the ozone layer is depleting. Farmers in the world have continued to use pesticides and fertilizers that contain harmful compounds that destroy the ozone layer. The continued use of methyl bromide ($CH_3Br$) in farms has been a key factor in ozone layer depletion.

$CH_3Br$ is an effective pesticide used to fumigate farm soils and various crops. Due to the fact that the pesticide contains bromine in its composition, it destroys the ozone layer having a depletion value of 0.6. The manufacture of this pesticide in many countries has been banned completely. Fertilizers containing nitrogenous compounds have also been considered as potential causes of the ozone layer depletion. Nitrogen compounds emitted in small amounts of NO, $NO_2$, and $N_2O$ have been classified as potential depletion factors of the ozone layer (Anderson, 2015). Many countries have encouraged farmers to use natural manure in their farms to improve soil fertility.

*Carbon tetrachloride ($CCl_4$):* Carbon tetrachloride ($CCl_4$) is another compound that is used by man which results in the depletion of the ozone layer. The compound is primarily used as a raw material in modern factories. This compound was also used as a solvent in many industrial processes. However, research conducted indicated that the compound was carcinogenic when used as a solvent. Industries refrained from using it as a solvent. Today, the compound is used to catalyze a certain process that requires the release of chlorine ions. The continued use of this compound has been a major challenge to the ozone layer because it has a depletion potential of 1.2 on the ozone layer (Anderson, 2015).

## 1.4.2 Effects of Ozone Depletion on Human Beings

Having understood the causes of the ozone layer depletion, it is wise to try and understand the potential effects and implications of the depletion on human beings and the plants and animals on the earth. This understanding is important because it provides knowledge of the dangers resulting from the ozone layer depletion.

**Table 1.1:** Effects of ozone depletion on human beings

| Effects | Characteristics |
|---|---|
| Agricultural food shortage | • Ozone layer depletion has resulted in shortage of food to human beings.<br>• The UV rays are altering the physiological and developmental processes in plants decreasing their productivity.<br>• Due to the fact that human consumption is largely dependent on crops for food, there is an increased chance of food shortage if the ozone layer is not protected. |

| | |
|---|---|
| Troubles on aquatic systems | • Ultraviolet radiation has greatly affected the aquatic ecosystem by resulting in the decrease of phytoplankton which is the basis of an aquatic food chain.<br>• The UNEP estimates that a 16% decrease in the ozone gas could lead to a 5% loss of phytoplankton, which could translate to a loss of 7 million tons of fish per annum.<br>• This loss would mean that the world would record a decreased consumption of proteins in human beings.<br>• This is because 30% of proteins consumed by human beings come from the sea and the percentage increases in the developing countries. |
| Damage to the Eyes | • Good eyesight is one of the best aspects of a healthy human being. However, in the recent development in technology, it is becoming a difficult problem to maintain healthy eyesight.<br>• This has been brought about by continuous exposure to UV rays.<br>• Direct exposure to these rays can result in cataract problems and snow blindness also known as photokeratitis.<br>• Cataracts are the world's major cause of blindness. If the ozone layer decreases by 1% this would translate to a 0.3% - 0.6% increase in blindness cases.<br>• These rays damage the eye lenses and cornea because they produce oxidative agents. |
| Skin Cancer | • Depletion of the ozone layer has been associated with the increase in the rate of skin cancer cases.<br>• The skin cancer that is most common is called non-melanoma which develops from continuous exposure to UV-B radiation for a long time.<br>• Statistics by the United Nations Environment Program (UNEP) show that an annual rate of 10% ozone layer depletion over a couple of decades there will be a dramatic increase in the skin cancer patients to 250,000 per annum. |

| | |
|---|---|
| Plants and Animals | • In some animal species increased UV-B radiation enhances the development of skin cancer.<br>• Skin cancer is a common problem in several animal species.<br>• This condition has previously been observed in cows, goats, dogs, sheep and lab animals.<br>• Skin abnormalities and infections in cattle can be blamed on the higher UV-B radiation.<br>• The radiation from sunlight reduces the growth of trees, changes the flowering pattern, initiates production of harmful substances and makes the plant vulnerable to infections.<br>• If this persists the species of the plant and the biodiversity can be lost. |
| Aging skin | • Continuous exposure to UV rays is harmful to our skins not only by causing skin cancer but also accelerates the skin aging process.<br>• Direct UV light rays penetrating the skin make the skin to look older than it actually is.<br>• In line with this statement, an individual suffering from this condition will look older than his/her actual age.<br>• Continuous exposure to UV light rays may lead to photo-allergy resulting in the formation of rashes in people with fair skin. |
| | |
| The Immune System | • The body defense mechanism that fights pathogens and disease-causing micro-organisms will depend on the strength of the immune system.<br>• Continuous exposure to UV light lowers the functioning capability of the immune system.<br>• Apart from the infections that affect the skin, the dangerous UV light affects the immune system of the entire body.<br>• Furthermore, continuous exposure to UV-B radiation will enhance the body to house and accommodate diseases instead of fighting them.<br>• In line with this statement, vaccination programs in the industrialized and developing countries could be useless and ineffective. |

**Source:** (Lucas et al, 2015)

## 1.5 GLOBAL ENERGY DEMAND AND IMPACTS OF GLOBAL WARMING

The supply of energy is essential to the sustained growth of the world economy. Because of the rapid growth of the world population, there has been a continued increase in the utilization of energy to improve the standard of living, or even to maintain it at its present level. However, energy production and utilization processes have contributed a lot to the menace of environmental pollution. A recent report indicates that present levels of consumption not only does adds to the problems of availability and cost of fossil fuels, but it is also a problem to environmental pollution. Even if ample supplies of fossil fuels could be assured, a time might come when increased utilization of fossil fuels cannot be allowed because of adverse effects on the environment, such as global warming and climate change (Haldar, 2010).

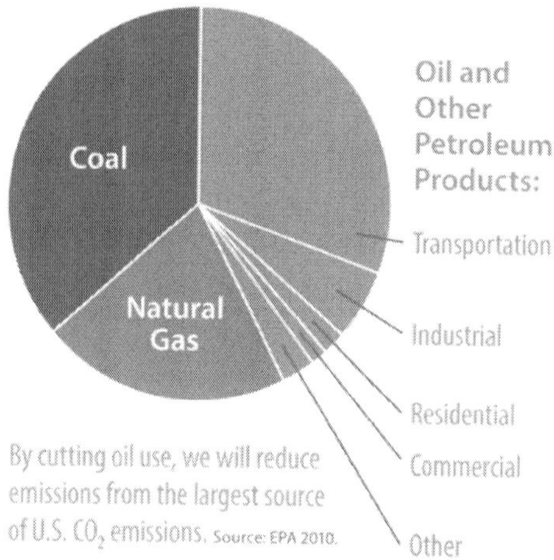


**Figure 1.2:** Fossil fuel is the leading cause of global warming.

**Source:** https://www.ucsusa.org/clean-vehicles/fuel-efficiency/benefits-of-reducing-us-oil-use

One can argue that the use of appropriate pollution reduction technology will reduce the environmental impact of fossil fuels. Pollution reduction by its very nature, however, creates additional energy demand and thus compounds the problems of availability and cost of fossil fuels. The general realization of the limited nature of fossil, fuel resources, as well as the environmental and safety constraints, has led to an examination of the possibility of utilizing those energy resources which are renewable. Renewable energy systems should be given a high priority as soon as possible. They have a critically important role to play in alleviating the impacts of global warming and climate change. Currently, new and renewable sources of energy account for approximately 15% of world commercial and non-commercial energy consumption (XinRong and Ibrahim, 2016). The only sources utilized to any appreciable extent are:

- Hydropower (1.7%)
- Firewood and charcoal (12 to 13%)
- Biomass conversion (0.8%)
- Tar sands (0.15%)
- Geothermal power, draught animals, peat, oil shale, solar power, wind power, and tidal power

Wave power and thermal gradient of the sea are essentially not utilized. The development of renewable energy resources is most urgent for developing countries with no reserves of conventional energy sources. These countries continue to import increasing amounts of commercial energy, especially oil. Persistent problems, with regard to the balance of payments, afflict these countries mainly because the price of the imported oil (and other fossil fuels) increases alarmingly while the prices of the products exported by these countries exhibit a tendency to decrease or fluctuate (Haldar, 2010).

## 1.5.1 Petroleum Products

The economic development of many countries around the world is heavily dependent on petroleum products. In 2010, for instance, petroleum products accounted for 85% of commercial energy consumption in Africa. The consumption of petroleum products by various consumer categories is outlined as follows.

- Transport sector. It comprises of different consumer units such as road, rail, marine and air transport. The transport sector is the major consumer of petroleum products. On average, this sector consumes 60-70% of all the petroleum products sold in the economy.
- Industrial and commercial sectors: These two sectors consume roughly 20% of petroleum products.
- Agriculture, power generation and government altogether consume about 15%.

In terms of quantity demanded, motor gasoline, jet fuel, light diesel, heavy, diesel and fuel oil are the most important fuels. Light diesel oil constitutes about 17% of total fuel sold every year. There has been a rapid growth in light diesel consumption mainly because of its use in the rapidly growing transportation sector whose services are required in the movement of freight and persons and also in boilers in the industrial sector. In many countries, commercial energy is mainly used by transport and industrial sectors. These sectors have been growing very fast and as a consequence, their demand for commercial energy has been rising. Their energy needs are primarily met by petroleum. This implies the position of petroleum products as a source of energy is not likely to change because at present and presumably the near future, there are no viable close substitutes for these fuels to be used by their major consumers (Weaver, 2011).

Increased commercial energy consumption can be attributed to the current high population and urbanization growth which entail the need for more energy for lighting, cooking and public transport especially in the urban areas. Furthermore, rural-urban dichotomy, which is a characteristic of most developing countries, and the localization of many industries in a few urban areas, necessitates rural-urban movements of, especially urban workers who are constantly in touch with their rural relatives since they are heavily dependent on.

Efforts to reduce this dependence on liquid fuels, which are a major source of global warming, continue and this is reflected in the gradual reduction in the share of imported fuels in the total commercial energy consumed (Weaver, 2011).

### 1.5.2 Shale Gas Development

Shale gas is naturally found in compact rocks that are less permeable and lies beneath the surface. It is trapped in extensive layers on rocks in every continent. According to Thomas et al. (2), the Obama government believed that its extraction would have help to reduce the impacts of global warming and climate change. It caused natural gas production to increase by 20% in 2010, and the government predicted a 46% percent production by 2035. Engineers have found two different technologies to extract natural oil while considering the environmental effect of their processes. Shale gas is extracted through Hydraulic fracturing and horizontal drilling. Hydraulic fracturing technology involves the injection of chemicals into the inner layers of the soil. Because it is natural gas, shale gas extraction has increased across the globe.

Shale gas extraction using hydraulic fracturing is safer than the extraction of minerals such as coal, which poses a threat to workers in the coal mines as well as negative environmental impacts (XinRong and Ibrahim, 2016). To create pressure in the shale rock, fracturing fluid is injected into the rock to create pressure and release hydrocarbon. In hydraulic fracturing, the source rock is made more permeable to cause the gas to flow to the well. Hydraulic fracturing opens the source rock to a distance of around one hundred meters. The micro-fractures created by the hydraulic fracturing results from pressure from water, chemical additives, and sand. The three substances form part of the fracturing fluid. A very high-pressure mixture of the fracturing fluid is injected into the well through the appropriate casing. The pressure water then creates a path for the gas through the cracks. Sand holds the cracks up while the chemicals protect the reservoir from bacterial contamination from the surface. Other substances used include viscosifiers, which ensure that the sand is suspended in water while the cracks are open.

Shale gas extraction using hydraulic fracturing poses little challenges to the workers at the wells since its effects cannot be felt immediately on the site. The fracturing liquid is injected using pipes and water tanks with large trucks are driven to the site. However, the environmental impacts of hydraulic fracturing make it an unsafe practice.

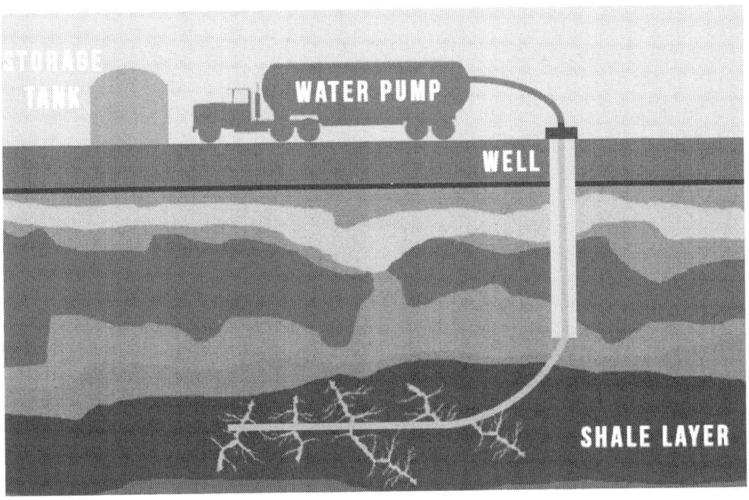


**Figure 1.3:** Hydraulic fracturing.

**Source:** https://www.wikihow.com/Better-Understand-Fracking

The following are potential impacts of hydraulic fracturing.

*Leakage:* Leakages are common in wells used to extract shale gas causing methane to migrate under the earth. Extraction of shale gas impacts on the environment by leaking both waste and extraction chemicals into the water suppliers. During extraction, greenhouse gases also leak causing pollution and improper processing of the shale gas. The challenge of reducing the leakages is that the natural gas extraction varies between different wells even in the same project. A strategy used to reduce the leakage in one well may not be enough in other wells within the same project. Despite the release of methane through leakages, hydraulic fracturing of shale results in fewer emissions of greenhouse gas as compared to fossil fuels.

Methane, however, has a global warming effect despite its infinite presence. The cement used in the wells normally becomes weak and the casing either fades away or becomes obsolete. There are poor maintenance and safety practices in the wells, which also leads to leakages. Air emissions are effects that take place during the extraction. It is primarily the escape of methane from the well and the different equipment used during the extraction process. Drilling for the shale gas using hydraulic fracturing results in smaller leaks of greenhouse methane gas than the federal government had estimated. The total amount of leaked methane gas is more than one million tons in the US annually.

*Radioactive waste:* Shale contains a great level of radium. This chemical element is resourceful in cancer treatment and industrial x-ray diagnostics. It is less required in oil fields. The fraction process displaces radium-tinged water that flows through the well up to the soil surface. Radiation levels can pile up at the of pipeline scale, the bottom of the tanks and the substance exposed to wastewater. The radium is displaced to the surface when water cracks the formation and hydrocarbons are released. Even the spent filters that are used in straining radium –laden create disposal problems since they contain elevated levels of radiation. There are regulations put in place to control illegal disposal especially in the states that experience intense fractionation processes.

*Drinking water contamination:* Water is an essential element of human growth and development in all living things, which is why its contamination threatens the life of both animals and plants. One of the problems faced by the oil drilling companies that use hydraulic fracturing is the contamination of water during the extraction process. Contamination of the surface water happens through spillage and cases of improperly built waste pits. Other reasons for contamination may include mistakes done during the drilling process or just poor maintenance mechanism.

The waste pits are in some cases not well-maintained causing contamination of the surface water. Fluids are not supposed to escape the fractured formation, although this happens in most cases. When the fluid escapes the fractured formation, especially in abandoned wells, the groundwater becomes contaminated. Hydraulic fracturing also requires the use of a large amount of water. The waste cannot be used for drinking while a lot of water is also used and cannot be recycled for drinking. This excessive use of water poses a danger of climate change and human health effects in society today. While international bodies join to protect water sources and conserve water, oil company focus on drilling and extraction of shale gas using a large amount of water. *Earthquakes:* Micro-earthquakes, also known as induced seismicity can also result from hydraulic fracturing. Micro-seismic activities help to mark the vertical and horizontal extent of the fracturing. Although the magnitude of the micro-seismic activities is too small and does not reach the surface, tremors resulting from fluid injection have been felt by people on several occasions. One of the causes of a magnitude 3 earthquake in the US is the water pushed towards the fault tends. The water causes earthquakes through a fault or slippage. Smaller earthquakes were detected between January 2011 and February 2012. The small earthquakes were over 109 and most of them resulted from fluid injection creating pressure at the preexisting sub-surface faults next to the wellbore.

## 1.6 CONSEQUENCES OF GLOBAL WARMING

Humanity may face serious economic consequences for the failure to address global warming. The melting of ice caps, the rise in sea levels and floods can destroy farmlands leading to widespread hunger or deaths. The need for fuel, wood for charcoal and other necessary products derived from deforestation, one of most harmful practices of mankind. It supplies people with ample goods, it helps to sustain people on a daily basis, but what are the consequences of such an action?

- The atmosphere would be contaminated with large amounts of carbon dioxide,
- There would be a decrease in rainfall, and
- When rain actually falls, soil erosion can occur.

- In addition, natural habitats are destroyed and there is a lot of damage done to valuable sinks for environmental pollutants.

The burning of fossil fuels increases the amount of carbon dioxide in the atmosphere, like its copartner mentioned above and decreases the amount of oxygen in the atmosphere. Similarly, the use of automobiles, carbon emissions from the burning of gasoline to power cars, trucks, and other methods of transportation also add to the pollution of the atmosphere. They form a great threat to humans as they produce a harmful chemical called carbon monoxide. Once humans interact too closely with it, it forms a stable compound with hemoglobin in blood which does not allow the efficient transport of oxygen (Haldar, 2010).

The mining of oil, coal and other mineral products allow greenhouse gases from the earth to escape, which would disturb the soil's stored gases and give them free movement to the earth's surface. The greenhouse effect is the natural warming of the atmosphere when certain gases such as carbon dioxide absorb extra energy from the sun. There is however an extreme to this effect; there is an enhanced greenhouse effect which is additional warming due to gases that have been released into the atmosphere as a result of human activity.

The constant disposal of waste into the sea disrupts the stable food chains formed by aquatic organisms. Plankton which is one of the most abundant primary producers is slowly diminishing because of the increase in temperature and other elements of climate change. The reduction of this prime producer can diminish the marine life in total, since the chain would be primarily disrupted. In reference to the exact causes of the depletion of effective marine life, humans and their family chains are also adding to the degradation of the environment. The increase of the earth's population comes with an increased demand for food, housing, clothing and all the necessities of life and guess what; they all have an impact, whether direct or indirect, on the devastating effects of global warming (Haldar, 2010).

Different countries around the world have suffered severe drought. Most of these drought conditions are normally succeeded by very wet conditions which culminate in very severe flooding in most parts of the world. Countries are plunged into a long famine resulting from the interplay between two extremes of drought and floods. In some parts of the world, these unusual weather conditions recur with even greater intensity during the long rains. The widespread flooding causes lake levels to rise dramatically. Rivers burst through their banks flooding low-lying areas. Many bridges are destroyed or submerged for periods of days and many roads and rail embankments washed away. River gauging stations are destroyed, equipment washed away and some channels suffer so much erosion that the stations have to be abandoned. Meteorological departments attribute these unusual heavy rains to the effect of global warming and climate change been experienced in different parts of the world. The most peculiar feature of the rainfall during the episode is its continuous and widespread nature. On the recurrence interval of the floods, the rains result from unusual meteorological conditions whose recurrence periods are beyond the present day knowledge to predict (Haldar, 2010).

The Technological Industry is directly impacted by all those issues stated above. The industry seeks to improve the severe impacts of global warming but bring along some harm to its practices. It is quite hilarious how mankind destroys an element that destructs the environment but then finds ways and means of reproducing and using that same element in some other form to do the exact thing it was made to do in the first place. It has been found that green plants are the only feasible way to remove carbon dioxide from the atmosphere. Solar energy is used by plants to fix the carbon content of the atmosphere into sugars which partially replace the fossil oil and gasoline. In addition, modern technology has found a way to lower greenhouse gas concentrations.

- Fossil fuels can be burnt with little or no restrictions
- Restrict the use of fossil fuel so that most of the coal, oil, and gas are left in the ground
- Introducing a technology that would allow us the use of our fossil fuels without emitting Carbon Dioxide ($CO_2$). Thanks to modern day technology it has already been introduced; it is called, 'Carbon Dioxide Capture and Storage or CCS.

## 1.7 RELATIONSHIP BETWEEN GLOBAL WARMING AND BUSHFIRE

The presence of wide forest areas densely populated with implantation of natural plants provides a combustible element that works as the fuel to generate the incident of bushfires. The spreading time of bushfire is the quickest and fastest. Due to this, human lives and environment face a certain loss. According to the research of (Anderson, 2015), these bushfire incidents occur 10,000 times every year more specifically in the areas of Australian states.

During the spread of fire, substances are incorporated with each other to elevate the reaction and accelerate the spread speed. One good example is the relative humidity. The presence of lower relative humidity value results in the absorption of water molecules due to which combustion reaction easily proceed by increasing the spread rate and wind transfer the fire from one place to another. As the temperature increases the combustion process increases and expedites the spread rate. The wind speed has an intensive effect as the wind speed increases to lower the relative humidity and accelerating the fire spread more vigorously (Anderson, 2015).

It is necessary to mention that the frequency, modeling and bushfire behavior approach is indicating the prediction according to which the alternative years other than drought years, there is an alteration in the cycle of bush fire. Usually, the dry season is the most sensitive to bush fire.

### 1.7.1 Fire Behavior Models

- *Surface Fire*: The surface fire spreads due to the contiguous fuel of dead and lives connected with ground. It is estimated as the initial and the most significant technique of spreading grass and bushfires. The given fire spread model has related the vegetation characteristics with the fire characteristics that are supposed to burn in that vegetation. This aspect is particularly for the spreading of surface fire and forecasts the spreading rate towards the way of maximum spread, the fire intensity, and flaming height.
- *Crown fire*: The crown fire behavior is the model based on tree tops. Crown fire depends on the distance of the height to crown fuel based (tree tops) and the presence of the crown fuel moisture substances. Also, three different crown fires are observable in the shape of active, passive and independent fire and exhibited with the same property according to which the fire threshold intensity is the contingent with the bulk intensity of crown.
- *Spot fire*: Spot fire appears at the head of fire front in the form of ashes incorporated with air and deteriorating downward from the wind fire. The spotting height spread from the fire front is considerably small, the spot fire location surpasses through fire front and the effect of this fire is reversed on the spread rate. However, the occurrence of breaching the fire break and spotting is also considered according to which the fire break appears as the region of low-slung rather zero fuel. Ashes can

appear substantially more ahead from the fire that results in the production of a new fire also required to considered and modeled individually from the core forefront but the probability is still low and also depend on the fire intensity.

## 1.8 EFFECTS OF SUN IONIZING RADIATION DUE TO GLOBAL WARMING

There are three possible categories of harmful effects of sun ionizing radiation.

- Transformation of somatic cells leading to the induction of malignancy in the subjects exposed
- Mutations in reproductive cells, with the potential for inducing genetic ill-health
- Injury to the proliferating and differentiating cells and tissues of the embryo or fetus leading to malformations and developmental abnormalities

The International Commission on Radiological Protection (ICRP) has classified radiation effects as either stochastic or non-stochastic. Stochastic effects are those for which the probability of an effect occurring rather than its severity is regarded as a function of dose, without threshold. The possibility that some radiation effects may be stochastic in nature implies that there may be no dose which is low enough to be regarded as completely safe. This realization forms the basis of what has come to be known as the ***ALARA Principle*** in radiation protection, which stipulates that all unnecessary exposure to radiation should be avoided and that the doses received during necessary exposures should be kept "As Low As Reasonably Achievable (ALARA)". On the other hand, the non-stochastic effects are those for which the severity of the effect varies with the dose, and for which there may be a threshold dose below which the effect is not observed (Haldar, 2010).

Carcinogenesis is now believed to be the major somatic risk of low-level exposure to sun ionizing radiation. Evidence that radiation produces cancer in man is obtained from epidemiological studies among groups of people who received high doses of radiation. Numerical estimates of the risks of carcinogenesis induced by low doses of radiation are obtained by extrapolation from observations at high doses. Such estimates are highly dependent on the dose-effect relationships assumed in the extrapolation. Most risk estimates in radiological protection are derived by linear extrapolation, assuming direct proportionality between dose and effect, although for radiation of low linear energy transfer (LET), such as x-rays, the linear-quadratic dose-effect model is found to fit a large number of experimental observations (Haldar, 2010).

Studies using white populations show a positive association between incidence those living at lower sea levels and mortality from melanoma. 'There is sufficient evidence in humans for the carcinogenicity of solar radiation. Solar radiation causes cutaneous malignant melanoma and non-melanocytic skin cancer. There is limited evidence in humans for the carcinogenicity of exposure to ultraviolet (UV) radiation from sunlamps and sun beds': IARC (1992); (Parkin et al, 2011).

The epidemiological problems encountered in low-dose mortality studies include:

- The very large sample sizes for both irradiated and control populations that are required to establish statistical significance at high confidence levels
- The complicating effect of the usually long latent period between exposure and cancer induction (in view of the non-specificity of radiation carcinogenesis)
- The uncertainties surrounding the possible synergistic role played by various physical, chemical and biological agents which may interact with radiation to induce cancer

In view of these problems, the evidence adduced in support of increased cancer incidence from direct observations at low doses must currently be regarded as inadequate. Quantitative estimates for the risks of carcinogenesis in various organs and tissues have been made by authoritative expert organizations. These estimates show relatively high induction risks for the female breast during reproductive life, hematopoietic tissue the thyroid gland, and the lung. The mortality risk from female breast cancer from low-LET (linear energy transfer ) radiation is estimated as $2.5 \times 10^{-3}$ $Gy^{-1}$, while for leukemia and thyroid cancer the risks are $2 \times 10^{-3}$ $Gy^{-1}$ each.

In addition, leukemogenesis shows a generally shorter latent period than the solid malignancies. Incidence risks are higher than the mortality risks, typically by a factor of about 2. The total mortality risk from all cancers averaged over both sexes and all age groups is estimated as being about $1.2 \times 10^{-2}$ $Gy^{-1}$. Irradiation of the embryo and fetus has been associated with an increased risk of childhood malignancies. Epidemiological studies among the Japanese A-bomb survivors have been the most important single source of human data on the late effects of radiation. Individual doses to these survivors have for a long time been based on the tentative 1965 dose (T65D). But in 1976, the United States Department of Energy released some previously-classified information which necessitated a re-evaluation of dose estimates. The re-evaluation and the potential effects of the revision on risk factors obtained from this important source of information are still being studied. Some experts have predicted that risk factors for low-LET radiation are unlikely to be altered by more than a factor of about 2 (Haldar, 2010).

Genetic effects may be induced by radiation through gene mutations or chromosomal damage. Among the harmful effects of genetic origin are:

- Stillbirths
- Mental retardation
- Mongolism
- Various malformations

Radiation is only one of the many agents with the potential for causing genetic harm, and it is considered to be a minor contributor to genetic disorders in humans. There is hardly any evidence of radiation-induced genetic effects from direct observations in man. Most of the data used to predict genetic harm in man are, therefore, derived from animal experiments, particularly in mice.

Several important observations and inferences have been made from these laboratory investigations:

- The frequency of radiation-induced mutations depends on the radiation dose, the dose rate, dose fractionation, and the time interval between irradiation and conception.
- The sensitivities of different mutations vary considerably. Any quantitative evaluations of sensitivity, therefore, involve averaging procedures.
- The radiation dose required to double the spontaneous rate of mutations is about 1 Gy (range 0.5 - 2.5 Gy) per generation, for radiation delivered at low dose rates, or in low doses. At genetic equilibrium, this level of exposure would induce genetic effects in about 10% of all live-born offspring in man.
- The male spermatogonia are much more sensitive to radiation-induced mutations than the female oocytes.

For purposes of assessing individual distress, the risk of serious hereditary ill-health in children and grandchildren following the irradiation of either parent is taken to be about $10^{-2}$ $Gy^{-1}$, and the additional risk to later generations to be the same.

On the assumption that only 40% of dose on the gonads is likely to be genetically significant, an average risk factor of $4 \times 10^{-3}$ $Gy^{-1}$ has been recommended for hereditary effects, as expressed in the first two generations. This risk factor is about one-third the total risk for carcinogenesis. Radiation-induced genetic effects are now considered to present a lesser degree of risk to populations than previously thought. The assessment of the potential genetic burden in a population from a given source of radiation exposure is based on an estimate of the genetically significant dose (GSD). This is defined as that dose of radiation which if it were received by each member of the population, would produce the same total genetic injury to the population as do the actual doses received by the individuals in that population. The GSD may be interpreted to represent the average dose to the gametes that will be effective for reproduction. Studies of the GSD from x-ray diagnosis in many different countries show values well below those that would cause concern. The estimated values of annual GSD, typically in the range 0.1 - 0.5 mGy, are lower than the estimated mutation doubling dose by a factor of several thousand times (Haldar, 2010).

Exposure of the embryo or fetus *in utero* may result in:

- Prenatal death
- Malformations
- Growth defects
- Death in the neonatal and infant periods
- Increased risk of cancer developing during childhood

These effects have been observed in both experimental animals and man. The observed effects of prenatal exposure depend strongly on:

- The gestational age at the time of irradiation
- The total radiation dose
- The dose rate

In man, during the first 2 weeks post-conception the main effect is the death of the conceptus. The period of major organogenesis, extending from the 3rd week to about the 10th -week post-conception is associated with a high risk for malformations and developmental abnormalities. Sensitivity to a particular organ malformation varies with the stage of development, being highest during the relevant stage of organogenesis.

Observed effects of embryonic exposure include:

- Growth retardation
- Microcephaly
- Hydrocephaly
- Mental retardation
- Co-ordination defects
- Other teratogenic effects

In man, the most important effect appears to be mal-development of the brain. Irradiation during the fetal stage (60 days to term) is associated mainly with growth retardation and the possibility of cancer induction in early life. Some evidence suggests that there may be increased sensitivity to mental retardation between weeks 8 and 15 of gestation. Exposure of the fetal oocytes may result in irreversible cell loss and consequently to reduced fertility of the offspring later in life. The risks of teratogenic effects are reduced during the fetal stage.

The probability of radiation effects on the embryo and fetus decreases with decreasing dose and dose rate. Pre-natal death, congenital abnormalities, growth deficiencies, and mental retardation appear to be non-stochastic effects. For most effects, experimental evidence suggests a threshold dose of the order of 0.1 Gy, but some animal studies have reported positive observations at lower doses. The whole question of whether or not thresholds exist for radiation effects arising from prenatal exposure remains unresolved at the present time.

## 1.9 URBANIZATION AND GLOBAL WARMING

Urbanization is the increase in the number of people living in towns, which is commonly caused by industrialization and people moving in search of better standards of living. Urbanization has consequences which are both adverse and beneficial as they have an impact on social and environmental areas. Temperature increase is a major problem brought about by urbanization this in turn, is brought about by the cutting down of vegetation in order to bring up residential structures and industries to cater for the increasing urban population. Temperature increases due to increased absorption of the energy from the sun and production of heat in the industries and the increasing human activities (Ariel and Robert, 2011). Air pollution has been impacted adversely by urbanization. Factories and automobiles which are common in the urban centers emit harmful gases and smoke hence resulting in air pollution. This gases and emissions are a major contribution to the problems of global warming which is becoming a huge health hazard. Increase in urbanization led to changes in the water cycle. This is because cities have more precipitation than the surrounding environments.Natural resources are destroyed in order to create space for the construction of houses. These forests that are cut down are also the natural habitats of animals and birds. Because of deforestation, the absorption of carbon dioxide is reduced, thus, increasing the impacts of global warming (Scott, 2016; Miller and Hutchins, 2017).

**Figure 1.4:** An illustration of how urbanization has contributed to global warming.

**Source:** https://www.newsecuritybeat.org/2012/07/urbanization-and-the-global-climate-dilemma-2/

## 1.10 REEVALUATING THE COAL POLLUTION

Global warming and climate change issues have always provoked immense attention and arguments from varied quarters. It is well known that modern times have seen an extraordinary increase in the average global temperatures, which have been attributed to various environmental conditions like:

- Unpredictable weather patterns
- Devastating natural catastrophes such as droughts and famines
- Floods
- Wild fires

Global warming has been attributed to increased globalization which has resulted in the generation of immense proportions of carbon dioxide into the atmosphere. The overwhelming effects of global warming have stressed the need for the enactment of legislation and acts that would limit the amount of carbon dioxide generated. Unfortunately, such legislation has to be a balancing act between the economic benefits pertaining to industrial processes and the need to protect the environment. This is the controversy that was evident in the recent Supreme Court ruling that upheld the EPA's authority to control the smog that comes from coal plants and drifts beyond state lines (IPCC, 2014).

**Figure 1.5:** Burning coal is the leading cause of global warming in the world.

**Source:** https://www.rt.com/business/global-power-cost-energy-522/

The main issue with the case is the pollution that is spread by coal plants. The EPA has been anticipated to propose fundamental new Clean Air Act regulation to reduce the release of carbon, which is termed as the greenhouse gas that traps heat between the ozone layer and the surface of the earth thereby cumulating the average temperatures. In the United States, coal plants are the largest generators of greenhouse gas emissions. In the case presented to the Supreme Court, it was noted that the soot and smog coming from the coal plants were blown to the east coast, thereby resulting in numerous ailments including lung diseases, asthma, as well as premature deaths. It should be noted that the efforts of the coal plants had only taken into consideration the welfare of upwind states while neglecting the downwind ones. Of course, the concern of the government is based on the need to safeguard the health and wellbeing of the people in all states and ensure that the benefits that are generated by the coal plants do not compromise the wellbeing of citizens or even the environment itself (XinRong and Ibrahim, 2016).

The general public would be concerned about the quality of air that they breathe. As noted, the smog and soot that the coal plants produce is connected to premature death, asthma and lung ailments. Further, they would be concerned with the high levels of carbon dioxide, which have been responsible for global warming. In essence, they pronounce that there should be more stringent controls on the amounts that are produced. In the case of the government/EPA, its main concern is the quality of air within the states. It should be noted that the agency strives to ensure that every other American has clean air to breath. In essence, it discourses that the operations of the industry would need to be based on sound science and strong legal foundations. Unfortunately, this is not the case as the health of citizens and the environment of the downwind states has been compromised by the pollution of the coal plants and the inadequacy of the laws governing such operations.

The coal plants are primarily concerned with their bottom-line or rather the profits that are gathered in from the production of electricity. It should be noted that the requirement for the reduction of carbon emissions and the putting of more safeguards to protect the downwind states would have a negative effect on the plants' profits. Making a decision pertaining to the most appropriate course of action presents numerous challenges as a result of the numerous competing interests of the stakeholders. In essence, borrowing from ethical reasoning theories would be imperative. The course of action could also have a bearing on the rates of employment in the states. It should be noted that adherence to the regulation could be so costly to them that they would have to close down their dirtiest and oldest plants. Not only would jobs be affected but access to affordable energy would be hampered in the states (XinRong and Ibrahim, 2016).

While crushing the ruling could go a long way in sustaining the employment rates and access to affordable energy, it should be noted that the costs are way too high compared to giving EPA the authority to control the amounts of emissions. Indeed, the enactment of the regulations does not necessarily mean closing down the plants but rather it could give the plants an incentive to come up with innovative ways for producing electricity or burning coal without producing as much carbon dioxide or at least protecting the citizens and the environment from the harmful effects of the same.

While there may be differing opinions, upholding the authority of EPA and allowing it to control the emissions and operations of the coal plants is the most appropriate way to go. Not only is it just and fair, but it has the highest benefits for the country at large and states themselves. Further, it safeguards the human rights of the citizens of the area, as well as the employees who can be assumed to be living close by and, therefore, under the risk for contracting ailments from the carbon dioxide emissions. Of particular note is the fact that the regulations would not necessarily cause business entities to close rather they could be inventive and adopt other cheaper and cleaner technologies. Preventing the EPA from taking control would be unethical given that only the business entities reap profits while making the lives of the constituents of these states miserable. Already, diseases and deaths have been reported as a result of the failure of the coal plants to incorporate the standards laid out (XinRong and Ibrahim, 2016).

## 1.11 THE BURDEN OF GLOBAL WARMING

The world is facing the impacts of global warming such as water scarcity, loss of diversity, desertification and sea level rise. Global warming has made most countries in the world experience high humidity, high temperature and arid conditions which have resulted in land damage and soil degradation. The intrusion of salt into aquifers has threatened food security in the world because it has affected agricultural lands. In the last few decades, most countries have observed a drastic change in rainfall pattern and noticeable dust storms (Anderson, 2015).

There is a need for the development of a strategic plan for adaptation and mitigation of global warming in the world. Stakeholders should be engaged to create an adaptive capacity to boost food security in the face of global warming and climate change. Adaptation of global warming and climate change also requires various stakeholders to alter their behavior. It is impossible for research to improve an adaptive capacity when a comprehensive understanding of decision making and the ability of the decision-makers to change is not available. The adaptation process is usually constrained by the political, social, institutional and economic environment which humans operate. The world is both biologically and physically threatened by the phenomenon of global warming (Bhupinder et al, 2011).

Several research studies have helped to gain insight tipping points that are possible in pointing to major changes in the current policies and farming systems that are needed, due to greater adaptive capacity and awareness of stakeholder views. These studies have also helped in better understanding of constraints and decision processes related to adaptive capacities. Still, they have helped in the setting of a collaborative process for stakeholders to come up with research agendas for developing policies and adaptation decisions options.

However, there is a need for more research studies on global warming and climate changes in order develop a learning process that is collaborative to support the food and agricultural systems adaptation to keep up with the various impacts attributed to these conditions. Such studies should mainly focus on:

- Proper indicators of resilience for use to global warming and climate changes by several decision makers
- Creating and evaluating ways for engaging the agricultural stakeholders in decision making and adaptive management at different scales on global warming and climate change adaptation
- Coming up with mechanisms to create awareness and education on global warming for adaptation and mitigation among individuals, institutions, and governments
- Evaluating and determining how information on global warming can be mainstreamed best in decision making

## 1.12 WHO SHOULD BEAR THE BURDEN OF FIGHTING GLOBAL WARMING?

Global warming is one of the current trending global issues. Since the industrial revolution and drastic human civilization, there has been a dramatic increase in activities that cause climate change around the world. These activities include:

- The release of carbon into the atmosphere through running of various automotive engines

- Deforestation
- The release of a large number of aerosols

Though global warming is also as a result of natural phenomenon such as volcanic eruptions, human activities can be managed and controlled to reduce the effect of climate change (Rakesh et al, 2014). However, the question of who to bear the burden of curbing global warming has become a problem. Some have argued that this responsibility of curbing global warming should be upon the developed countries. This view is however subject to objections. There arc three storms that need to be addressed when deciding who to bear the burden of curbing global warming. These are:

- Global storm
- Intergeneration storm
- Theoretical storm

Countries should contribute to the climate change managing cost in proportion to their emission of greenhouse gases. This approach is also referred to as 'contribution to problem' and is seemingly convincing as the developed countries are the one that has emitted most of the greenhouse gases (GHGs) that are causing global warming and therefore should contribute a significant share of the managing cost (Rakesh et al, 2014).

Other facts supporting the idea are that:

- Developed countries are also resourceful hence they are in a position to meet the cost
- One ton of the GHG has a similar effect irrespective of where it is emitted
- Scientific methods have been developed to measure the exact release of the gases and its source. This can help in equal sharing of the cost among the nations.

*Global storm:* This storm emerges from the three most prominent characteristics of the climate problem. These features are presented in the following table:

**Table 1.2:** Features of global storm

| Features | Characteristics |
|---|---|
| Fragmentation of Agency | Responding to global warming becomes a major challenge to humanity since institutions, and specific individuals that are the agent's causes are not unified by a common platform.<br>Global governance is not centralized, and therefore it becomes challenging to propose and implement responses to global warming.<br>Some of the institutions available that attempts to address the issues of global warming such as the United Nations faces problems of lack of full global membership, strategies to divide responsibilities to the member nations, inadequate funding among other.<br>The ineffectiveness of global governance makes it difficult to structure and implement agreements related to measures of fighting global warming. |
| Dispersion of causes and effects | Emission of the greenhouses gases which is one of the primary causes of global warming is a worldwide phenomenon.<br>These gases can be released from any point in the world where human activities leading to the release are carried out.<br>The gases move to the atmosphere and play a role in impacting on climatic patterns around the world.<br>Identifying who will take the responsibility of addressing issues of global warming becomes a problem since it is difficult to pinpoint the sources of the greenhouse gases.<br>The effect on climate is also not only felt at the origin of these gases, but it spreads far wide around the globe. |
| Institutional inadequacy | There is universal agreement that outlines the need to change the structures in existence when the conditions are favorable.<br>Doing this will help in solving issues on global warming with all parties involved in the process.<br>These structures can be changed by introducing a system with enforceable sanctions.<br>Proposing this may seem simple in theory, but its application in practical situation faces many challenges.<br>Some of the challenges include the absence of an efficient global governance unit and lack of unification of institutions that contribute to climate change. |

**Source:** https://en.wikipedia.org/wiki/Effects_of_global_warming

Twisted liabilities on the issue of global warming interact mostly with global power and influence. The highest contributors of elements leading to global warming are:

- The developed
- The wealthy
- The most powerful nations around the world

Nevertheless, the effects of global warming are deeply felt by poor and developing countries since they have limited resources to counter attack these effects. The other problem is that they are not in a position to hold the developed countries responsible for their actions due to their little bargaining power and influence on global issues. The view to burden the issue of global warming to developed countries also face challenges as the developed countries gives attention to issues they term of more moral concerns such as human rights violation, peace, and global poverty. This results in less attention and abandonment of the problem of global warming (Rakesh et al, 2014).

A global agreement on how the responsibility to curbing global warming should be shared faces other challenges such as:

- Uncertainty in the magnitude of distribution of emissions and effects
- Unproven scientific data on the amount and impact of emissions from a particular nation
- The possibility of trust issues arising on the costs charged on each country towards the management of global warming
- The idea by some nations that the world is a better place with global warming than without also pose a challenge of collaboration. These nations argue that global warming has come along other major world revolution in terms of industries development and infrastructure that have improved the lives of billions of people around the world. The industrial revolution, for example, a period that is associated mainly with huge emissions of greenhouse gases into the atmosphere, brought about the development of many industries and considerable improvement in technology.

*Intergenerational storm:* The main objective of this storm is the idea of burdening developed countries with the responsibility of global warming management in that the phenomenon of human-induced global warming is a slow process that takes a very long time to be realized (Eastwood, 2011).

Some effects of global warming due to greenhouse gases such as rises in sea level will be felt after an extended period of the release of the gases. There is a fact that a molecule of carbon dioxide gas, the highest component of greenhouse gases, can spend between 5 to 200 years in the atmosphere. This period is long enough to bring about a lagging effect. Furthermore, global warming is a very robust phenomenon and therefore the actions to reduce the carbon dioxide present in the atmosphere need to be advance-planned. The effects and evidence of global warming being experienced currently are as a result of emissions dating back many years ago hence determining the source and amount from a particular nation become a major challenge.

Because of the back-loading nature of global warming, it becomes harder for connecting causes with the effects. Since the impacts of climate change being felt now are as a result of emissions from many years ago, there set in the problem of historical responsibility problem. Morally, those countries that released the emissions in the past cannot be condemned since the effects were not yet studied. It, therefore, pose a challenging issue of who should assume the cost of managing the effects already experienced. The lagging impact of climate change results in delayed action hence more severe repercussions in the ability to manage the issue. Moreover, the institutions tasked with the responsibility of managing climate change may face challenges since they have relatively short time horizons.

*Theoretical Storm:* Theoretically, this is the third objecting view of placing the burden of addressing global warming on developed countries. Many global warming theories and effects face various difficulties in assessing even fundamental issues such as scientific uncertainty, intergenerational equity, and nature among others. These and other issues such as predicting the emission contribution of different nations using theories faces objections on their accuracy hence can be dismissed if used to assign responsibilities to different countries. The theoretical storm, combined with Global and Intergenerational storms give rise to the problem of ethical response on climate change (Eastwood, 2011).

Researchers have recognized the evolution of unequal emissions among nations. In other words, most of the developed countries have already consumed a big part of the atmosphere that has the capability of absorbing carbon dioxide and other greenhouse gases. One can argue that if all the largest contributors to the emission of greenhouse gases into the atmosphere decided to announce a radical cut of emissions, this act would reduce the risks to future generations through the decreased possibility of catastrophic effects. If such a declaration were made, it would raise many ethical and moral questions. This is because many countries' economies are laid on a foundation of industries mostly run by fossil fuels that emit a lot of carbon into the atmosphere. Cutting these emissions would mean that most of these industries would shut down resulting in loss of a source of livelihood to many and eventually massive effects on the economy of the country. Even with the need to secure the future generations from effects of global warming, such a move may lead to social and economic catastrophes. The global system cannot cope with such a drastic declaration, and the result can be unending social problems (Eastwood, 2011).

Final note: This chapter discussed the basic idea behind the concept of global warming along with its effects. The aspects of the effects of ionizing radiations were also discussed. An interesting aspect has been covered in this book as to how to handle this issue and the storms or challenges of who should bear the brunt of this phenomenon.

# Global Warming, Water, and Agricultural Crop Yields

# 2

## CONTENTS


2.1 Introduction .......... 36

2.2 Soil Water Availability and Crop Farming .......... 36

2.3 Global Warming Impacts on Rainfall Availability For Agricultural Practices .......... 38

2.4 Agro-Climatic Zones .......... 40

2.5 Surface Water Hydrology .......... 41

2.6 The Impact of Global Warming on Water Chemistry .......... 43

2.7 Plant Distribution and Water Relations .......... 45

2.8 Soil Water Availability .......... 47

2.9 Importance of Plant Water Relations To Production .......... 47

2.10 Soil Moisture Stress on Photosynthesis .......... 52

## 2.1 INTRODUCTION

Different crops have varying degrees of tolerance to the effects of global warming in relation to their survival and yield reduction. It has been recognized that for crop water use studies are to be meaningful, while the yield response to the crop to water deficits must be known. Knowledge of the yield and crop water use is needed to predict the level of production that can be achieved by the use of the available water resource.

There are several approaches used to assess the effects of water supply on agricultural productivity with reference to different effects of global warming such as drought. These approaches include:

- Rainfall either as seasonal total or at particular stages of crop development
- Drought indices, based on the absence or deficiency of rain can also be computed and correlated with crop yield
- Computed soil water status values for assessing the effect of water supply on yield
- The use of water loss from crop surfaces as an index of crop yield and growth

Climatically, computed values of potential evapotranspiration have also been used to assess the effects of global warming on potential crop yields. For most crops, a linear relationship exists between seasonal water use and dry matter production. The slope of this linear relationship varies with each crop. This relationship is valid only when other growth factors such as fertilizers, temperature, sunshine, and soil depth are not limiting with good crop management practices. For most crops, dry matter production does not determine the actual yield. This depends on the harvested part of the plant.

## 2.2 SOIL WATER AVAILABILITY AND CROP FARMING

The amount of water available to the plants during the growing period can be analyzed using the water budget. This enables farmers to assess the actual period over which plants can grow with adequate moisture available for their various physiological processes.

The water balance equation is represented below:

$P=E_o+S+R_o$

Where,

P represents precipitation

$E_o$ represents the evaporation

S represents the change in soil water storage capacity

$R_o$ represents the runoff including the percolation

It is important to always bear in mind that this formula is used where there is no inflow such as irrigation. This water budget formula is also not applicable in areas where irrigation is rarely found, the analysis may not deviate from the actual environmental conditions that a small-holder farmer faces. Mostly, the information contained in the water budget can be used to establish the water availability to show the conditions graphically. Since the rainwater percolates into the ground where it is stored (as the water reserve) the length of the growing period is not limited to the end of the rainy season. Therefore, depending on the water holding capacity of the soil, the period in which water is available to plants is always longer than the actual rainy season. When the soil water reserves are considered to be available for plant development, humid and the sub-humid periods are experienced throughout the year (Bhupinder et al, 2011).

In some places, the main climatic constraint is a low temperature which often leads to occasional frosts. This factor selectively limits the range of crops that are found within these places. A look at the available data (rainfall, evaporation, and temperature) and checking them against the various climatic or agro-climatic classification systems, these places can be related to:

- Tropical rain forest, permanently humid with a short dry season
- Upper highlands with permanent cropping possibilities divisible into a long to very long cropping
- Cool tropics or tropical highlands with a growing period of 270 days

These climatic conditions allow for the growth- of peas, potatoes, cabbages, maize (only in frost-free lower areas) and pyrethrum. Besides crop farming, animal husbandry, mainly dairy cattle and sheep-keeping, is a widespread economic activity. This is made possible by the cool climate and availability of high-quality feed from grass and Napier grass. However, these climatic conditions do not provide sufficient bases for planning for agricultural development in these areas. They show the degree to which global warming effects and climatic elements could be a source of great potential and constraints to plant growth (Bhupinder et al, 2011).

**Figure 2.1:** Global warming effects and climatic elements are the main sources of great potential and constraints to plant growth.

**Source:** https://ccafs.cgiar.org/blog/climate-smart-soil-water-and-nutrient-management-options-semi-arid-west-africa#.W_-h54szbcs

## 2.3 GLOBAL WARMING IMPACTS ON RAINFALL AVAILABILITY FOR AGRICULTURAL PRACTICES

One of the most essential environmental factors that influence agricultural practices is the availability of rainfall. There are different effects of rainfall on agricultural practices.

- Rainfall determines the potential of a place in terms of the crops which can be grown
- It determines the timing and sequence of farming operations
- Rainfall is also an important element in terms of supplying soil moisture. The proportion of rainwater that will constitute the soil moisture reserve depends largely on rainfall and soil characteristics that include: the total rainfall, rainfall distribution, rainfall intensity and duration, soil depth and moisture holding characteristics, and the balance between rainfall and evapotranspiration.

The seasonal occurrence of rainfall in different parts of the world normally controls the extent to which the agricultural rhythms occur. There are several global warming and climatic conditions that influence the seasonal occurrence of rainfall. These include:

- The general circulation: Different parts of the earth are influenced by different trade winds. For instance, the eastern part of Africa is influenced by the South-East and North-East Trade winds: In the middle part of the year, an incursion of westerly winds occurs that affects mainly the western sector. The seasonal variations of rainfall are the result of the seasonal shifts of the Inter-Tropical Convergence Zone (ITCZ) which is related to the apparent movements of the sun.
- Topography and exposure: The effect of topography and exposure is mainly on the amount of rainfall received at a place. Topography acts to make the ITCZ rather diffuse.
- Latitude: This determines the length of the rainy seasons and the timing of rainfall maxima and minima.

Although it is difficult to give precise limits of the onset, duration, and cessation of the rainy seasons, a general picture should be given because it is important from the farming community's point of view. An examination of the seasonal rainfall totals may indicate that adequate amounts of rainfall for a particular crop were received. This may, in reality not be so, for it is important to know the frequency of occurrence of rain and no-rain periods of different durations. Apart from revealing the pattern of the rainfall generating mechanisms that operate over an area, the study of wet and dry spells yields data that can be helpful in agricultural planning, especially with regard to planting, supplementary irrigation, erosion control strategies, the control of forest fires, and use of machinery on the farm (Hakeem, 2015).

It should be appreciated, however, that even if the seasonal rainfall regimes are objectively determined and delineated and that data are available of the wet and dry spells, soil and crop management factors should be taken into account to further define the availability of the growing season. As a first approximation, it is useful to know the seasonal rainfall regimes that occur in an area, and in addition, some rainfall characteristics in each regime. The rainfall characteristics which are of special agricultural significance include the occurrence of wet and dry spells of various lengths.

Radiation from the sun is important from the point of view of photosynthesis and evaporation. The amount of solar radiation reaching the surface of the earth is a function of the following.

- Seasonal variations of the atmospheric conditions of dust and water content
- The type and amount of cloud cover
- The extinction coefficient of the clear sky
- The earth-sun geometrical relationships
- The slope/aspect of the surface

Temperature is another very important element attributed to global warming and climate change. This is because of its influence on evaporation and on plant growth and development. Since air and water temperatures are largely dependent upon solar radiation, a fairly close correlation between them and the rate of evaporation can be expected. The temperature of the water surface is important in that it governs the rate at which water molecules leave the ground and enter the overlying air. A change on water surface temperature may, therefore, have a profound short-term effect upon the rate of evaporation. Ultimately, however, it is the net movement of water vapor molecules, and the way in which the water molecules leaving the water surface are redistributed within the lower layers of the overlying air, which largely conditions the evaporation process (Casper, 2010).

Apart from these chains of reactions, temperature is very important in weathering processes, as they are largely responsible for contraction and expansion of rocks which lead to the physical breakdown of a rock mass into relatively small fragments. For instance, research has shown that that evaporation decreases by about 1% for every 1% increase in salinity, so that evaporation from the water in soil with an average salinity of about 3.5% is in some cases 2-3% less than evaporation from other surface water. This reduction is brought about by the reduced vapor pressure of the saline water. However, the case of non-marine water, salinity values are normally so small that they can be discounted when comparing evaporation rates of different places. Atmospheric humidity, on the other hand, is important because of its effect on evaporation and also on the incidence and virulence of plant diseases and pests (Eastwood, 2011).

## 2.4 AGRO-CLIMATIC ZONES

Agro-climatic zones normally offer the first impression of the effect of global warming in a certain place. This determines the suitability of various land use alternatives with emphasis on the suitability for particular crop varieties. Agricultural crop farming in regions like Africa depends on rainfall. The potential for rain-fed agriculture varies from "very high" in the zone I to "medium" in zone IV (Anderson, 2015).

Apart from defining and indicating the spatial expression of the seasonal rainfall regimes occurring in an area, it is important for practical purposes, to study the rainfall characteristics in each regime. A rainfall characteristic of great importance to water resources utilization in general and to agriculture, in particular, is the occurrence and duration of wet and dry spells. Research work along these lines in many parts of the world is lacking. Therefore, an examination of rainfall seasonality in an agricultural area is required. This is more so given the divergent results of the various investigators regarding the onset, cessation and consequently the duration of the rainy seasons.

The interpretation of water components characteristics reflects the spatial variation of global warming and climate change in a place. Therefore, it is important to carry out a systematized regional climatic description based mainly on the rainfall and temperature characteristics. On the basis of soils and climate (rainfall and temperature) and their integration, agro-climate zones are determined which determine areas better suited to the various crops and agricultural land-use (Bhupinder et al, 2011). In relation to the agro-climate zones, certain tree species may be suited especially in terms of meeting multiple needs such as browse or fodder soil conservation, fencing and fruit or horticultural needs. Some trees have good potential in vegetation cover systems.

## 2.5 SURFACE WATER HYDROLOGY

Hydrology is the science of water over and under the surface of the earth. There are different components that comprise hydrology. These include:

- Stream flow
- Water vapor
- Precipitation
- Soil moisture
- Groundwater

Explaining the hydrology of a place is very significant because it provides information on the various hydrological components which may throw light onto the state of groundwater resources of the place and the factors governing its occurrence, distribution and quality. In most cases, there is too little water for use by human beings while sometimes there is too much. Some regions have abundant lakes and rivers while other areas are practically devoid of water. Global warming is one of the major factors that govern the hydrological cycle. This has often been attributed to various climatic conditions such as flooding and drought. An uneven distribution of water over the earth may be attributed to these factors governing the hydrologic cycle (Ouda, 2015).

Apart from global warming and climate change, there are other geological factors such as soils and vegetation that are also known to influence the distribution of water. However, it is mainly the weather and climate that form and sustain lakes and rivers, create deserts and produce floods and droughts. Different places around the world experience a wide variation in climate due to great differences in altitude. Along the coastline lies a narrow tropical belt behind which stretch large areas of semi-arid and arid conditions. In regions experiencing extreme climatic conditions such as the Middle East, water is often in short supply.

Most lake basins receive good rainfall. For the efficient development of water resources, it is useful to know the total amount of water available for various uses. The variability of this amount is almost as significant as its average value. The water resources of most lake basins consist of many perennial rivers and streams. For efficient planning of river and stream water, the annual regime is quite important for the discharge of the rivers and streams. The variability is mainly due to rainfall. Any particular year may have one or two periods of high water at the time of long and short rains. Occasional years have no high water season, where both the long and short rain spells fail due to global warming and climate change impacts. It is during the periods of low flow that volume of water in a river becomes critical for water supply in agricultural irrigation.

Due to extraction and evaporation losses, discharge of the rivers and streams further down-stream may be considerably less. At the end of a very dry season, river flow may be sustained only by the spring flows, but at other times the flow recovers increments for a certain distance along the river channel derived from groundwater bodies downstream of the mainsprings. When the channel passes out of such groundwater bodies, the flow begins to decrease. The point at which this discharge occurs is known as the point of maximum flow, though this point may vary according to the season. Losses of flow downstream of the point of maximum flow are greater in the lower reaches of the rivers. A statistical analysis of the river flow over a long period offers estimates of the quantities of water available for certain percentages of time. A compromise must be adopted between the two extremes when estimating water availability-on an all the year-round basis (Ouda, 2015).

## 2.6 THE IMPACT OF GLOBAL WARMING ON WATER CHEMISTRY

The effects of global warming, which vary in different places of the world result in a diversity of weathering products which in turn influence the nature and amount of solutes in the water. These solutes are reflected in water chemistry. It must, however, be appreciated that global warming and climate change have not always remained the same. The effects of past climates, although modified by the present, still persist in paleosols and old erosion surfaces occurring in arid and semi-arid areas.

Temperature is strongly influenced by the altitude of the land surface and, to some extent; the same is true of rainfall, for higher areas tend to have a consistently high rainfall while the rainfall regime in the low areas ranges from very low to high. Temperature affects the rate of weathering which tends to be more rapid with higher temperatures. Rainfall affects the soil by leaching; the higher the rainfall, the greater the leeching of soluble products of weathering from the soil. Both rainfall and temperature, therefore, affect the weathering of the underlying rock. Intense weathering is closely associated with high rainfall areas. It has been noticed, for example, that in areas of recent volcanic ash, intensive weathering has taken place where rainfall is higher, whereas less intense weathering is associated with less rainfall. Rainfall intensity is quite important in the rates of weathering. The higher the intensity of rainfall, the greater will be the erosive power of the rain. This is particularly true of the drier areas where the vegetative cover tends to be poor. Although the total amount of rainfall may be low, appreciable erosion may occur if the intensity is high (Daniel & Cynthia, 2011).

From an agricultural perspective, the amount of rainfall, its variation in space and time, its reliability and its intensity are the rainfall elements which are most important to the farmer. These same elements are also of great importance in weathering and consequent provision of the chemical solutes in groundwater. Water soaking into the ground or being absorbed by the rock or soil (infiltration), has a profound effect on the chemistry of the surface and subsurface water and varies widely over a catchment area in response to a number of often closely related factors such as rainfall intensity, amount and rainfall chemistry.

It has been observed that if the rainfall intensity is lower than the infiltration capacity of a soil or rock, all the falling rain is not held on surface storage would infiltrate into the soil or rock. This then means that, there would be a direct relationship between the rate of infiltration and the intensity of the rainfall. When, however, the rainfall intensity exceeds the infiltration capacity, the above relationship breaks down and may indeed be replaced by an inverse relationship between infiltration and rainfall intensity. This is normally the case when an increase in rainfall intensity is reflected as an increase in size of the raindrops and consequently with increased compacting forces as the drops strike the ground surface (Rao et al, 2016).

Rainfall drops may compact the surface of the soil such that infiltration is rapidly and effectively reduced. This effect is more noticeable in clay soils which can be rendered virtually impermeable than on clear sandy soils which are much less susceptible to rain compatibility. It has been revealed that most of the fine particles that are normally carried in suspension by infiltrating water may clog pores in the soil surface and so lower the infiltration rates. The rate of infiltration may vary with the depth of standing water, particularly during the early stages of the infiltration process when only a thin layer of surface soil has been saturated so that the resistance to infiltrating water is comparatively low (Rao, et al, 2016).

### 2.6.1 Factors Influencing Water Infiltration in Soils

- One of the main factors governing the magnitude of infiltration is the time available for the infiltration process so that infiltration is generally greatest in conditions which slow down the movement of water over the ground surface. The nature of the surface is, henceforth, very important in this respect.
- Another factor influencing water infiltration into the soil is the quality of water. Most water passing through the soil surface collects fine clay and silt particles and carries these suspensions into the soil profile, where blocking of smaller pore spaces may occur. Infiltration rates have also been found to vary when the infiltrating water is contaminated by salts, particularly in very alkaline soils because the salts affect not only the viscosity of the water but also the rate of swelling colloids. Since the viscosity of the water and therefore the ease with which it may move through

the soil pore spaces, varies with water temperature, the latter tends to exert some influence upon infiltration rates.

- It is also known that infiltration rates for given precipitation conditions are generally higher during periods of higher temperatures than during lower temperature seasons.

## 2.7 PLANT DISTRIBUTION AND WATER RELATIONS

The presence or absence of a particular plant in a place is controlled by definable external and internal factors. The external factors include physical factors such as mineral salts, light, temperature, wind, precipitation and biotic factors such as influence of other organisms. The internal factors include the physiological, biochemical and anatomical factors inherent in a plant. All these factors, though many and complex, influence the patterns of plant distribution. Such distribution patterns could be described as random or uniform or clump distribution.

According to their photosynthetic mechanisms, it is believed that all green plants have the capability to fix carbon dioxide through what is known as the Calvin pathway of photosynthesis. In this process, carbon dioxide is assimilated reductively by the enzyme ribulose bisphosphate carboxylase and metabolized through the Calvin cycle to produce the first stable three-carbon compound phosphoglyceric acid. It is from molecule that the plants which exhibit only the Calvin pathway of carbon fixation derive their name of $C_3$ plants.

A number of genera of monocotyledons and dicotyledons contain both $C_3$ and $C_4$ species. The $C_4$ pathway appears to have evolved more recently probably in response to the increasing effects of global warming. This pathway is most common in the Gramineae family where about 40% of the species are $C_4$ plants. These $C_4$ plants possess distinct characteristics which may be used to differentiate them from $C_3$ species. These groups can be distinguished on the basis of their anatomy and chloroplast structure or position in the bundle sheath cells:

For a long time, there has been an expressed wish in the world to obtain a reliable way of classifying vegetation on the basis of their production potential based on available soil moisture. Additionally, such a classification needed to be based on criteria that could be observed directly. The first attempts to meet those needs were made by Thornthwaite in 1948, who proposed a moisture index based on the balance between rainfall and evaporative demands of the site. The index proved promising at medium and low altitudes where a low temperature does not limit plant growth. The system was based on the fact that the soil water available to plants is mainly controlled by the balance of water between the incoming rain and the evaporation losses. Therefore, Thornthwaite built his index on purely climatic factors which could be assessed with no reference to the vegetation types. The main drawback of the Thornthwaite system was that he estimated evaporative demands on the observations of temperature based on the correlation between mean monthly temperature and solar radiation (Casper, 2010).

Under temperature conditions, where both daily mean temperature and daily radiation depend mainly on day-length, the correlation was reasonable. However, in the tropics with near constant day-length, the daily temperature is not an adequate guide to radiation. Therefore although Thomthwaite's concepts were valid and useful they did not give an accurate estimation of available water, especially, at high altitudes.

The distribution of plants is generally discontinuous, meaning that there are always gaps between individuals and clumps occupied by individuals. Depending on the width of the gaps, plant species have two levels of distribution namely: (a) geographic distribution and (b) ecological distribution. On the geographic basis species distribution is indicated on the map by dots or generalized boundary lines. Within these large areas, the given species may only occur here and there under favorable habitat conditions. On the ecological basis, the distribution of a species is considered in terms of habitat potentials.

Different climatic and environmental features are used to delimit the distribution boundaries of a species. The vegetation of the world is made up of a wide range of vegetation types. But with the exception of alpine vegetation, all other types are mainly distributed in relation to the available soil moisture. At one extreme where water is in the least supply, one may find arid bushed grasslands and thickets while at the other extreme one finds closed rain forests and swamps (Casper, 2010).

Although rainfall has a big influence on the amount of water available to plants, due to a number of reasons, total rainfall is not as important as soil water balance in controlling plant distribution. One cause of a lack of correlation between mean rainfall and vegetation distribution is the very seasonal nature of tropical rains. Often large quantities of rain may fall in a short period of time leaving the rest of the year dry. Since the soil cannot hold more water than its physical nature can allow it to hold at field capacity, the excess water received either percolates into the bedrock or it is lost as runoff. The other reason which causes a lack of correlation between total rainfall and vegetation distribution is the variable nature of tropical rains. It is very common for a place to receive an annual amount of rain which is very different from the average expected. This can lead to different vegetation types characterizing places with the same annual average rainfall. To obtain a way which can relate some aspect of the physical environment to site ecological potential, methods such as the evapotranspiration formulae and the available moisture index have been designed (Casper, 2010).

## 2.8 SOIL WATER AVAILABILITY

There are several elements that determine the availability of water in the soil profile. These include:

- The depth of the soil profile: The soil depth of interest is the effective rooting zone of a crop. Different crops and varieties vary in their rooting depth.
- Soil water storage capacity: The amount of water stored and available to the crop is the difference between the field capacity and permanent wilting point. The amount of water stored by a soil depends on soil characteristics such as soil structure and texture.
- The moisture extraction range of the crop: A coarse-textured soil loses water quickly through rapid percolation, whereas finer soils and soils of high organic matter content retain water longer.

## 2.9 IMPORTANCE OF PLANT WATER RELATIONS TO PRODUCTION

There are numerous environmental factors that influence the growth and development of plants in terms of quality and quantity. Water balance is one of these factors, and it is normally affected by the effects of global warming in one way or the other. Though the growth and development of any plant would depend on its hereditary potentialities, the importance of water as a medium for the physiological process of plant growth cannot be overstressed.

Water plays a very crucial role in the growth and development of plants.

- It acts as a medium for nutrient absorption
- It is required for synthesis of metabolites and structural materials
- It is needed for transport of assimilates etcetera
- It is required for all important processes of plant growth and development

Therefore, a correct internal water balance is essential to the proper running and development of the plant system. Although the total water potential is not the only "property" to which plants respond, it is, however, an important property governing the direction of "source-sink" reactions in the plant system. When the water potential of the cell tissue or organ of a plant is zero (at zero it would be at equilibrium with pure free water) the individual cells are said to be turgid. However, if the water potential drops below zero levels, the turgor pressure of the cells also drops giving rise to flaccidity.

The condition of flaccidity is well shown by wilted leaves. Two main factors govern the turgidity or flaccidity of cells depending on which direction they offset the level of plant water potential. These are:

- The level of the soil water potential
- The diurnal lag of absorption behind transpiration

These factors are also influenced by both physiological and environmental factors. Absorption of water by the plant occurs because the suction at the leaves (or the water potential at any point in the plant system) is greater than the soil water potential near the roots. As the plant loses water through transpiration the leaf water potential drops. If this water is not regained by absorption, the leaf water potential will be progressively reduced. The cell turgor pressure declines rapidly as the leaf water content and leaf water potential are reduced and depending on the cell volume/turgor pressure relationships and the structural characteristics of the leaves concerned, they will gradually become wilted (Casper, 2010).

**Table 2.1:** Effects of the water balance on growth and development of plants

| Effects | Characteristics |
|---|---|
| Root development | • It is not yet clear how the relationship between water stress and root development occur.<br>• Some writers suggest that root development is enhanced relative to shoot development by water stress.<br>• There is a reason to expect that water stress would reduce elongation of roots due to a reduction in rates of meristematic activity directly associated with the level of internal water stress. |
| Photosynthesis | • When evapotranspiration demand exceeds water uptake, a condition of negative leaf hydrostatic pressure develops.<br>• A reduction of leaf turgor pressure induces stomatal closure.<br>• This creates a restriction in the supply of carbon dioxide necessary for photosynthesis.<br>• Severe stress may also affect photosynthesis through the dehydration of the photosynthetic apparatus.<br>• There is evidence to demonstrate that stomatal closure is the primary cause of the first stages of photosynthetic response to water stress. |
| Intake of nutrients | • The relationship between nutrient availability and plant water status is not a straight forward one.<br>• For instance, nutrient uptake is not directly related to water uptake, neither is nutrient availability a function of soil water imbalance<br>• The effect of senescence upon nitrogen and phosphorus levels in the leaves of tomato plants is confused with that of water stress.<br>• It has been observed that the nitrogen and phosphorous levels in leaves decrease with the increased water stress. However, considering young and old individual leaves, supplied with adequate water, it has been noted that the nitrogen and phosphorous levels decrease with leaf age, with the earliest and greatest reductions shown by the younger leaves.<br>• Therefore, an imbalance in nutrient supply arising from water stress does have a limiting effect upon certain aspects of plant growth and development.<br>• This imbalance should easily show up in the final crop yield. |

| | |
|---|---|
| Fordham effect | • Fordham, R., (1970) contended that yield is not necessarily a function of growth rate.<br>• He suggested that plucking itself would bring about uneven crop distribution.<br>• He suggested that since shoot elongation is merely sensitive to physiological drought than the initiation of primordial buds, buds are produced at a more constant rate than shoots.<br>• So long as drought is not severe enough to limit bud initiation, buds will continue to form and accumulate until the stress conditions are alleviated.<br>• Abrupt cessation of water stress conditions would be accompanied by the expansion and elongation of buds giving rise to a peak crop.<br>• In periods of water stress, however, few buds will develop into shoots and yield will be depressed. |
| Transportation | • When the water supply is abundant, only snail leaf water gradients would be required to transport water to the transpiring surface.<br>• Because the stomata remain open, and therefore the diffusive resistance due to stomatal impedance remains small, evapotranspiration is determined by environmental factors, in particular, those factors influencing air turbulence near the crop surface, and solar energy.<br>• As the leaf water content declines (due to desiccation or increased transpiration, so that leaf water potential is less than soil water potential in the root range), a level of leaf water potential is reached at which the stomata close.<br>• The events associated with stomatal closure include restriction of photosynthesis and the increase of leaf temperature due to the greater disposition of the absorbed energy into sensible heat transfer rather than evaporation.<br>• If this condition is protracted, lack of leaf turgidity would inhibit cell enlargement and subsequent growth.<br>• Still, diurnal increase in leaf temperature may induce midday closure by increasing the respiration rate and the level of intercellular carbon dioxide concentration. |
| Primordial Initiation | • Thc vegetative and productive primordial in the apical meristems are amongst the most sensitive parts of the plant to the conditions of water stress.<br>• Moderate bouts of water stress suspend the initiation of primordial. |

| | |
|---|---|
| Cell division and elongation | • Multiplication of plant cells has been shown to be at least suspended in cases where water stress has existed.<br>• Underdeveloped vegetation clearly manifests the effect of water stress upon cell enlargement.<br>• It is evident that in some species cell enlargement is so sensitive to water stress that stem elongation or leaf enlargement can be inhibited by small diurnal amounts of water tension that occur in plants normally enjoying plentiful water supplies in days of high radiation incidence.<br>• Lack of cell enlargement leads to poor organ development.<br>• Vegetative plants show signs of poor leaf development characterized by small surface area and therefore reduced photosynthetic activity and often limited net assimilation rate.<br>• This is indeed the first sign of indirect effects of soil water status on physiological processes having a key implication to final biological yield.<br>• In mature tissues, stress brings about the the migration of phosphorous from older leaves to the stems end meristematic tissue, and protein hydrolysis leading to the breakdown of normal cell function. |
| Protein Synthesis | • There is a close association between growth rate and protein synthesis.<br>• There is also a correspondence between protein synthesis and protoplasmic ingredients such as ribonucleic acid (RNA) content and between RNA and deoxyribonucleic acid (DNA).<br>• Because of the well-known effects of plant water stress upon plant growth, it is certain that water imbalance would have a limiting effect on transamination and subsequent protein manufacture.<br>• Water stress induces proteolysis.<br>• A reduction in the rate of increase of DNA with the imposition of slight stress levels has been observed in radish leaves<br>• Since nitrogen and phosphorus are essential components of the pyrimidine ring structures that constitute the nucleotide parts of DNA and RNA molecules, the suspension of primordial initiation can probably be explained by inhibition of DNA and RNA synthesis imposed by water stress. |

**Source:** (Lisar et al, 2012).

## 2.10 SOIL MOISTURE STRESS ON PHOTOSYNTHESIS

Water stress affects a wide range of plant metabolic processes. Water is involved in almost all biochemical processes of a plant and its reduction due to stress transduces a multiplicity of primary and secondary growth responses. Although a great deal of research has been carried out on this subject, more information is still required in order to discover all the links between water variations and their effects on the extent. Information is still required on the physiological events underlying the morphogenetic effects of water stress at different stages of particular crop plants. It is also necessary to know at which level of stress do processes such as photosynthesis begin to be markedly impeded in specific plants especially those exposed to stress.

Generally, the photosynthetic capacity decreases due to stomatal closure or due to the dehydration of the photosynthetic apparatus. When the stress becomes severe even negative values of photosynthesis may develop. The key compounds in the photosynthetic reduction of carbon dioxide remain unaffected until severe stress sets in. One of the commonly reported effects of water stress on carbohydrate metabolism is the increase of sucrose synthesis and an immediate suppression of polysaccharide synthesis. It is to this increase in sucrose synthesis that researchers have attributed the often reported reduction in starch levels due to water stress.

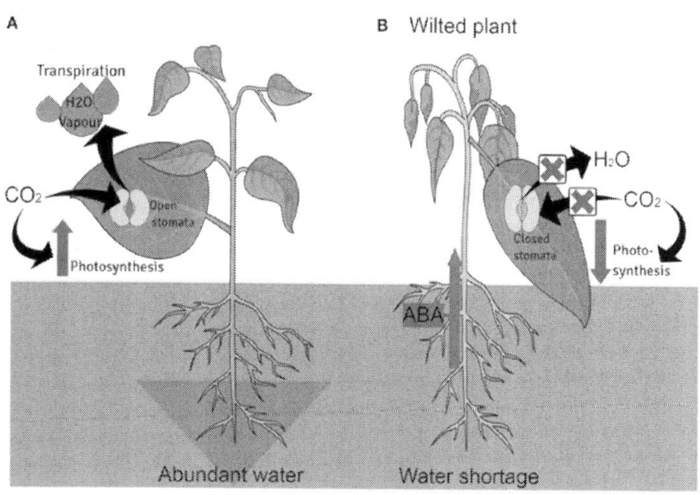


**Figure 2.2:** Effects of soil moisture stress on plant photosynthesis.

**Source:** https://kids.frontiersin.org/article/10.3389/frym.2017.00058

There are conflicting reports about the relationship between photosynthesis and adaption of plants to various effects of global warming. For instance, the pre-sowing hardening of plants causes drought-hardened plants to maintain a higher rate of photosynthesis than the controls during conditions of water stress. This increased photosynthetic rate is attributed to the higher hydrophilic nature of protoplasmic colloids which have been subjected to drought hardening treatment. However, it has been found out that there is no difference between the photosynthesis of the hardened and control plants under drought conditions. It would be interesting to find out whether the species naturally adapted to drought conditions as a result of global warming possess the photosynthetic features attributed to drought-hardened seedlings (Rakesh and Kalpana, 2014).

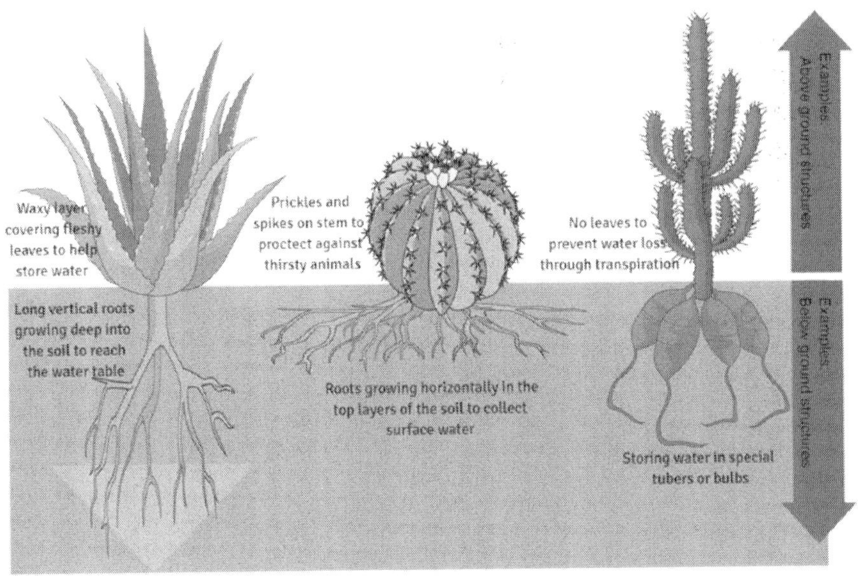


**Figure 2.3:** Structural adaptation in plants due to water shortage.

**Source:** https://kids.frontiersin.org/article/10.3389/frym.2017.00058

Although it is generally accepted that the major indirect influence of water stress on photosynthesis is on stomatal movement, Rakesh and Kalpana (2014) pointed out that stomata are insensitive to a reduction in plant water potential until a threshold is reached. It has been added that this threshold is between -7 and -18 bars depending on the type of plant species involved. In this case, one would not expect the stress influence mediated through the activity of stomata to proceed uniformly from the very beginning of the stress until its culmination. One would expect an initial period of no response followed by a gradual increase in response when the threshold is passed. It is necessary to know whether plants possess this kind of response.

Stomatal closure due to water stress does not only affect the uptake of carbon dioxide but also the rate at which moisture diffuses out of the leaf. The ease with which water vapor and gaseous exchange will take place between the mesophyll cells and the atmosphere outside the leaf will depend on the magnitude of resistances along the diffusion passage. Variations in leaf resistance might enable a plant to retain as much water as possible while at the same time being able to obtain enough carbon dioxide for photosynthesis. It has been suggested that stomatal oscillations may affect favorably the ratio between the assimilation of carbon dioxide and transpiration thus optimizing the relationship between assimilation and growth (Rakesh and Kalpana, 2014).

Final point: This chapter discussed the role of predicting rainfall in agricultural areas as well as the effect of global warming on water adaptations in plants, mainly crops.

# Environmental Sustainability, Global Warming, and Agricultural Land Degradation

# 3

CONTENTS


3.1 Introduction .......... 56

3.2 Three Fundamental Pillars on Social Development of Sustainable Development .......... 56

3.3 Actions to Ensure Environmental Sustainability .......... 58

3.4 Deforestation and Global Warming .......... 61

3.5 Resources Mismanagement and Global Warming .......... 63

3.6 Impacts of Population Growth Towards Environmental Degradation .65

3.7 Agricultural Practices and Environmental Degradation .......... 68

3.8 Global Warming and Agricultural Land Degradation .......... 69

3.9 Politics and Agricultural Land Degradation .......... 70

3.10 The Way Forward Towards Natural Resource Management .......... 71

## 3.1 INTRODUCTION

Throughout history, people have shown their capacities to exploit the environment to the point of severe environmental degradation. Presently, people live in a modern consumerist society where vast quantities of natural resources are consumed daily. 50 % more of natural resources are extracted by humans today than 30 years ago, at about 60 billion tons of raw materials per year (Overconsumption? Our use of the world´s natural resources. Sustainable Europe Research Institute (SERI), Austria and GLOBAL 2000 (Friends of the Earth Austria) – with: Institute for

Economic Structures Research (GWS), Germany; 2000). Sustainability focuses on a balance between the needs of society to develop and the importance of protecting the same environment to enable current and future generations to be able to enjoy the same resources.

There are three main elements must be observed in order to have a sustainable environment. These are:

- The function of natural systems
- How natural systems remain diverse
- The processes through which natural systems produce the needs of the ecology in order to remain in equilibrium

## 3.2 THREE FUNDAMENTAL PILLARS ON SOCIAL DEVELOPMENT OF SUSTAINABLE DEVELOPMENT

### 3.2.1 Environmental Protection

This involves the protection of ecosystems, integrity, and the regulation of air quality and resource sustainability with a focus on the factors that cause stress to the environment. It involves activities such as recycling and minimization of power consumption. Companies and industries are regulated in order to keep their carbon emissions at a minimum. Incentives are also given to individuals or companies that install renewable energy sources. The EPA has recognized the importance of technology in environmental protection and is keen to ensure the implementation of key technological advancements in its programs (White, 2012).

### 3.2.2 Economic Development

This mainly involves giving people access to their needs without degrading their quality of life and minimizing the financial stress that they incur. It is, however, not well defined due to the differences in political ideologies on economically sound decisions that affect people and the environment.

### 3.2.3 Social Development

Social development typically involves the awareness and passing of legislation that ensures the protection of agricultural land from pollution and activities that may be harmful. It also involves providing access to essential resources without degrading the quality of life. Finally, there is education which includes teaching the community the importance of environmental sustainability and benefits of environmental protection together with the dangers that come about due to lack of it.

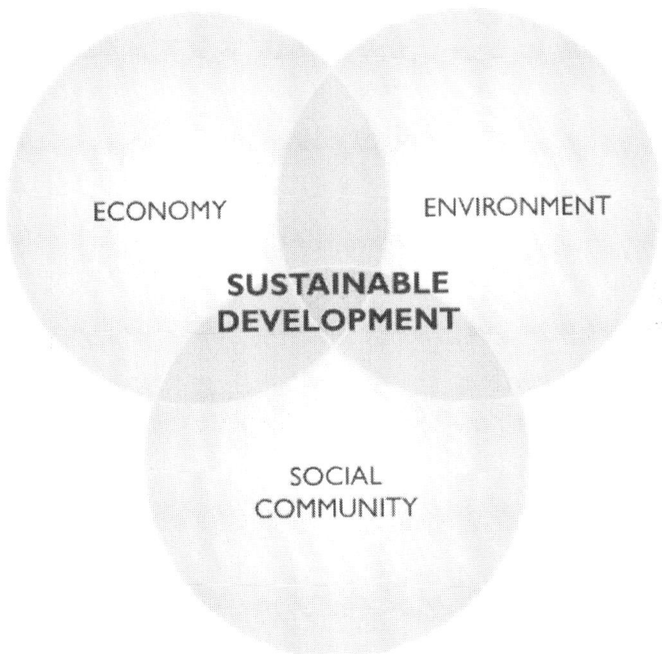


**Figure 3.1:** Fundamental pillars on social development of sustainable development.

**Source**: http://www.yourarticlelibrary.com/environment/what-is-the-importance-of-sustainable-development/9910

Environmental sustainability is the development of a responsible relationship between the people and the environment to reduce or totally avoid the depletion of natural resources and ensure environmental quality in the long term. Environmental sustainability ensures that the needs of today are attained without compromising the capability of future generations to exploit the same resources. The environment has a natural capacity to rejuvenate itself. A good example is seen in the natural life of plants whereby when they die, they fall, decompose and their nutrients are taken back by other plants that give rise to more plants through the natural process of reproduction. Nature alone can take care of itself, but when the people factor or the presence and activities of mankind is involved, then active measures have to be taken to prevent the depletion of resources (White, 2012).

## 3.3 ACTIONS TO ENSURE ENVIRONMENTAL SUSTAINABILITY

### 3.3.1 Sustainable Agriculture

This is the implementation of farming methods that protect the environment. Farmers can use less water, chemicals and pesticides. They can also apply better tillage practices such as zero tillage and crop rotation that ensures the soil retains its nutrients throughout the year.

### 3.3.2 Sustainable Energy

This is the use of renewable energy sources. It includes the tapping of energy sources such as wind power, tidal power, solar energy and biofuels. With improvements in technology, some of these energy sources such as solar energy which could not be tapped at a large scale can now be exploited for industrial application. The truth is that the traditional use of fossil fuels is under the risk of depletion and is the main source of greenhouse gases and other pollutants which harm the environment. However, the use of sources such as solar power and wind are sustainable means for future use and for mitigating pollution.

The organizations, especially the manufacturers, should incorporate renewable sources of energy like the wind and solar power. Here, a combination of these two sources will ensure that there is enough supply of power for running the firms' activities at the plant machinery level and the office equipment level. The government should formulate and implement regulatory policies that prompt organizations to use renewable sources of energy rather than contemporary sources such as wood and fossil fuel. The government should set policies that help subsidize the installment of windmills and solar panels for organizations. Other policies such as 'cut-one, plant-two' should be enacted for those who ought to use trees (wood) to ensure sustainability (Rattan et al, 2016).

The available must-use equipment like bulbs, boilers, etcetera, should be modified such that they can use as minimal energy as possible to avoid wastage. For example, energy saving bulbs should be used to cut on energy costs. Direct-sunlight boilers can be employed in place of conventional electric boilers to control power use and reduce wastage. Innovative modification of energy harvesters and users should be a priority for all corporations.

The machinery (such as car engines) that used fossil fuels can be dismantled and be used to create new equipment that does not entirely depend on fossil fuel. They can be re-modeled to use other sources of energy. Use of machinery such as automobiles that use excessive fuel should be reduced within the society and the corporate industry. Even better, the automobiles that consume fossil fuels should be replaced with those that use more biodegradable fuels such as ethanol. The production processes for organizations should be arranged such that they are in the top energy-saving and cost-effective topographies. Here, wait times in the production process should be mitigated to reduce the high demand for energy resulting from high wastage (Rattan et al, 2016).

### 3.3.3 Sustainable Forestry

This is the regulation of forest resources in order to cater to the needs of industries and communities while ensuring the forests remain healthy. This is done through activities such as selective logging which is the removal of particular trees while preserving the rest. Young trees can also be given time to mature before harvesting, trees can be planted through afforestation and protected forests can be created. Sustainability and sustainable development are responsibilities which fall on all the people. Government policies and environmental legislation may take people a mile but the greater journey lies with the individual initiatives that different people take. Humanity must develop an appreciation and love for the environment and a desire to ensure that future generations can enjoy the same beauty people have. Society needs a holistic and visionary approach to tackle sustainability. Normally, a narrow-minded or single -focused agenda may miss the mark significantly. People need to take advantage of the technological advancement, knowledge and skills that this generation has availed to people to preserve the world for future generations (Rattan et al, 2016).

### 3.3.4 Re-using and re-cycling Waste Products

The goods and services providers should use biodegradable packaging material instead of the traditionally used polluting materials. For instance, the use of polythene bags should be substituted with more biodegradable bags such as canvas or khaki materials. Otherwise, the packaging material should be possibly re-usable or recyclable. For example, old carpets can be re-processed into wall insulation materials (recycled), and drug holding containers can be used for keeping other elements such as salt or sugar (re-use) (POPLER). The waste produced within an organization can be transformed into usable and valuable goods. For example, the waste paper materials can be used in the art to create necklace and bracelet beads or to mold sculptures. The resulting beads and sculptures can be well decorated and sold to earn the organization more returns or be included as a social project involving the needy people within the society thus helping eradicate poverty (Rao et al, 2016).

**Figure 3.2:** Waste recycling and reusing.

**Source:** https://www.conserve-energy-future.com/importance_of_recycling.php

The personnel within an organization should be sensitized to avoid careless handling of waste and to participate in re-use and recycle art activities. Everyone within the organization should be engaged either directly or indirectly in ensuring that there is no dumping of waste within the firm and around them in society. Here, educating them on how to make extra cash through recycling and ensuring that they are conversant with the relevant environmental sustainability techniques is the best way for the organization to adopt the new strategy.

The corporation should make relevant policies to ensure that the personnel is obligated to the need for efforts of re-using and recycling. Requisite rules and regulations against unnecessary dumping should be implemented to regulate the waste. For instance, a fine can be applied to anyone who is found dumping specific waste such as plastic bags within the corporation. Also, reward-based policies should be implemented for the acknowledgment of the people who come up with the most innovative re-use and re-cycle propositions for managing waste, within and outside the considered firm (POPLER) (Rao et al, 2016).

As aforementioned, the waste should be used for utilitarian purposes which generate funds and other resources. For example, the plastic bottles can be used for making walls in developing countries for those without houses and the bottle-tops can be used for making carpets. The excessive wastes will be reduced in the environment if they are put to better use within society. An ultimate eradication of this waste can be done when the non-degradable materials are replaced with more biodegradable ones. Rather than making specific personnel responsible for waste management. Everyone should be made entirely responsible for waste management through the set policies. Hence, the management system should hold all people accountable rather than place it all on individual persons (Rao et al, 2016).

## 3.4 DEFORESTATION AND GLOBAL WARMING

Deforestation is a bad practice since it negatively affects the livelihoods of many people and massively threatens a wide range of animal and plant species. It involves the clearing of trees whereby land is later converted to other uses. Trees are often cut down and sold as fuel while the cleared land is used for plantation or used as pasture. Forests cover about 30% of the planet and the trees produce vital oxygen for people and wildlife (Pandi et al, 2010). Many people rely on benefits associated with forests such as fresh water, shelter, traditional medicine, food, clothing, and so on.

**Figure 3.3:** Current rate of deforestation is contributing a lot in terms of global warming.

**Source:** https://www.healththoroughfare.com/science/deforestation-will-increase-global-warming-more-than-expected-a-new-study-reveals/5468

Forests are under threat due to deforestation since the latter jeopardizes the benefits stated above. Forests play a vital role as relates to mitigating global warming since they act as a carbon sink (soaks up carbon dioxide, which would be free in the atmosphere and then contributes to changes in the climatic patterns). It is estimated that about 15% of all emissions of greenhouse gases are attributed to deforestation. In this sense, deforestation undermines the important function of the carbon sink.

Proponents of deforestation argue that this practice has an economic value since the trees that are cut down are often sold hence providing monetary value to individuals, and the bare land can then be used for other practices such as pasture (Pandi et al, 2010). However, the negative effects of this practice are much higher than this benefit as it hurts the livelihood and wellbeing of every living creature. Deforestation particularly affects rainforests, which is a major home for most of the world's biodiversity. Therefore, deforestation is harmful to every living creature since it limits the availability of oxygen to lives and may eventually bring about drought given that it affects the availability of rain. It is a major contributor to global climate change and should ,therefore, be discouraged at all cost, and for that matter, the government and other stakeholders should strive to combat illegal logging, promote certification of responsible management practices, protect forested areas, and reform trade policies.

## 3.5 RESOURCES MISMANAGEMENT AND GLOBAL WARMING

Excessive exploitation of natural resources such as forest has existed since early times. About 3000 years ago, China forests were nearly eliminated because of the clearance of land for agriculture, and for the provision of energy. The landscape was so devastated that annual floods and summer droughts became common. Bad policies and lack of effective management programs have led to the breakdown in the relationship between people and the environmental support systems. In the 1890 s to 1910 s, many families in the United States starved to death due to a drought on the Great Plains, but today rural people in industrialized countries survive such natural disasters because of soil and water conservation programs, crop insurance, rural credit, agricultural education and advice, forestry services, and fish and wildlife services (Sheldon & Michael, 2013).

The lack of good policies leads to degradation and depletion of greater resources. Environmental degradation in developing countries has emanated from the economic control of these countries by developed countries. About one-quarter of the population in the entire world which lives in the developed countries consumes three-quarters of the world's goods. The developed countries produce most of the goods and services but the raw materials are obtained from the developing countries. The chief economic activity in the developing countries is agricultural in nature, and from this theory of the developed countries over the developing countries it would follow that the developing countries would divert the best agricultural land to the production of cash crops for export to the developed countries (Sheldon & Michael, 2013).

Because of an increasing need for foreign currency developing countries have devoted more land to cash crops. This has led to the socio-economic causes of desertification, such as:

- Over-cultivation
- Overgrazing
- Woodland clearance
- Mismanagement of water resources

Such human activities lead to desertification which the United Nations Conference on Desertification (UNCOD 1977) defines as "the diminution, or destruction of the biological potential of the land, which leads ultimately to desert-like conditions and is an aspect of the widespread deterioration of ecosystems under the combined pressure of adverse and fluctuating climate and excessive exploitation" (Pandi et al, 2010).

The present-day environmental degradation in the developing countries had its genesis during the colonial era, and it was basically mismanagement of a fragile ecosystem due to the ignorance of the agricultural experts coming from the developed countries. In Africa traditional agricultural practices included:

- Shifting agriculture
- Forest fallow techniques
- Mixed cropping

However, colonial experts viewed this as "backward, feckless and irresponsible". Because of this, most of the African people were forced to abandon this practice and to adopt crop homogenization or monoculture, (patterned after the agricultural practices in the temperate countries), and to eliminate fallow systems. The experts had failed to understand that traditional farming practices were intended to preserve the environment.

The fertility of the tropical moist forests is not really in the soil but rather in the vegetation itself, and when the vegetation decomposes, (and this is quite rapid in a tropical climate), nutrients are released into the soil. This is not the case in a temperate forest system where a greater proportion of the nutrients are stored in the ground. In temperate climates monoculture could be practiced without interruption and without any need of a fallow period, because the soil is fertile, but in tropical countries where the trees and other vegetation hold the nutrients, clearing of the land leaves only a thin layer of nutrients which is quickly depleted or eroded during farming (Manjit & Banga, 2013).

Forest ecosystems display a remarkable resilience in the face of human intervention as long as the degree of exploitation is limited and do not exceed the threshold beyond which recovery is impossible. It has been noted that mixed cultivation, which provides specified diversity, and shifting cultivation enables the tropical environment to self-regulate and regenerate itself. Therefore, over-cultivation and excessive deforestation lead to reduced crop and tree production. These problems of deforestation and desertification have rendered the developing countries poorer, thus reducing their purchasing power of surplus products and services from the developed countries.

The policy by the developed countries to establish agencies to aid the developing countries in development, and thus reduce poverty while increasing the purchasing power has been ineffective. One of the objectives of this policy has been to enable the developing countries to have a more disposable income to meet the basic needs, such as domestic energy, and the necessary infrastructure. Unfortunately, the strategies adopted tend to negate any possibility of success in the realization of these objectives.

While much of the development aid is well-intentioned and humanitarian in nature, a lack of understanding of the social, cultural and religious factors of the recipient countries has tended to increase the problems. Governments in the developing countries do not consider desertification as a high-priority item and lack the political will to mobilize action in an attempt to correct it. Furthermore, the lack of research and demonstration capacity in developing countries has affected the degree to which mismanagement of the resources could be halted. It would appear that the effort to increase the disposable income if realized would provide the investment in technology that would be required in order to manage the resources efficiently. The argument that governments in the developing countries do not consider desertification as a high priority item seems exaggerated (Pandi et al, 2010).

## 3.6 IMPACTS OF POPULATION GROWTH TOWARDS ENVIRONMENTAL DEGRADATION

Rapid population growth rate, especially in the developing countries has been attributed to environmental degradation. This has led to the depletion of important agricultural resources such as land and water. The International Conference on Population in 1984 drew world attention to worsening situations such as:

- Expanding population size
- Depleting resources
- Intensifying underdevelopment
- Continuing environmental deterioration

These worsening situations were particularly directed towards developing countries. These situations have led to different agricultural land degradation such as:

- Arid and semi-arid formation
- Desertification
- The spread of the Sahelian conditions
- Devastating Drought

As a result of these conditions, the world is now unable to provide basic human needs of food, fuel, and shelter for large populations. This problem can be attributed to lack of a balance between people, resources, development and the environment leading to resource wastage and depletion, emanating from underdevelopment and poverty. These resources are:

- Land
- Forest
- Water

While over the last two decades many developing countries have achieved significant economic growth, such economic gain has been rapidly consumed by the additional population. The result has been:

- Increased poverty
- Food deficits
- Insufficient investment resources for increasing land productivity and employment
- Intensified pressure on land, water and forest resources

Because of the increased population pressure on the land, there is no longer any land available to be cultivated, and the ensuing over-cultivation leads to soil infertility and exposes it to erosion. Because of population pressure and poor management of the agricultural land, the rate of deforestation is ten times the rate of planting.

Fuel-wood is the main source of energy for some 2 billion people in developing countries around the world, and therefore, deforestation has diminished a valuable resource and increased human suffering. Because of the rising pressure of population, the land is becoming scarcer through sub-division for inheritance. Population growth and uncontrolled exploitation of natural resources combine in a vicious circle: the more the people there are the greater their destruction of the fragile environment, and the poorer this leaves them and their descendants (Keshav and Joshi, 2013).

**Figure 3.4:** High population growth rate has led to a high rate of deforestation.

**Source:** http://erdo-mezo.hu/2016/03/08/unios-eljaras-romaniaval-szemben-az-erdoirtas-miatt/

The reason why population growth leads to resource depletion and environmental deterioration is that people being rational, seek to maximize their personal gains over the common good. Each herdsman will try to keep as many cattle on the commons as possible, and that because of tribal wars, poaching and disease, the number of animals is kept below the carrying capacity of the land. As the population increases, each herdsman continues to increase his herd without any limit in a world that is limited. As each follows his own best interest, in a society that believes in the freedom of the commons, overgrazing follows and produces erosion and wind dominance. In general, a commonly owned environment, such as air or water is misused because nobody owns it, in other words, it is a "commons". Renewable resources are seen as the "commons" and because the majority of people have no real sense of ownership of these resources, then they fail to treat them with care (Spellman, 2014).

It has already been noted that traditional farming methods are discouraged by agricultural experts, and to a great extent, this knowledge has been lost. It has also been pointed out that in response to population growth land has been subdivided to uneconomical units leading to over-exploitation of the land. Reduced productivity of the land has been followed by underdevelopment and poverty. In such an environment of ignorance and hardship, people appear to be trapped in vicious circles so that, even if they wished to maximize the common good they do not appear to have the right means.

## 3.7 AGRICULTURAL PRACTICES AND ENVIRONMENTAL DEGRADATION

Humans play a great role in global warming. Since early times, they have used fire to modify the surrounding. Deforestation is one element that has contributed greatly to global warming. Due to human activity modification to the character of biomes, including the introduction of secondary forests, savannas, and desert margins has been prevalent. Introduction of alien species around the world is another disturbing factor. While farming is a rewarding job, it can cause negative environmental effects as well. Toxic gases from fermented crops and animal waste and pesticides are some of the hazards that pollute the environment and contribute to global warming. A wide variety of agricultural practices around the world determine the level of global warming. Ultimately, this is dependent on the production practices and system used by farmers (Christopher et al, 2016).

Agriculture plays a major role in the economy of society. Different agricultural methods, such as improper irrigation and inappropriate fertilizer use as well as other activities like deforestation and overgrazing affect land, air and water. The use of pesticides and other aerosol help in dealing with different pests hence increasing yield. This, however, can lead to ozone layer depletion, and thus global warming. Usage of fertilizers can be harmful to the air when evaporation occurs on the soil, thus leading to global warming if it occurs continuously.

Another key point is that much of the local natural vegetation covers like trees have been cleared for more crop production. Trees help in ground cover and act as wind breakers; therefore, they protect other weak plants from harmful effects of the elements. Trees also hold the soil together with their strong roots, therefore, controlling soil erosion. Clearing vegetation cover without planting new flora, results in more soil erosion because rainfall washes away fertilizer filled soil to the rivers affecting fish and other aquatic creatures. Another factor is overgrazing. Overgrazing is practiced on open land previously used for growing crops. Livestock grazing in large numbers tread on the surface soil, loosen it and cause it to be carried away by runoff water or the wind (Ariel & Robert, 2011).

From a farmer's point of view, crop production is a source of food and economic lively hood. Crops such as corn, wheat among many others, act as a source of food for both livestock and humans. On the other hand, the sale of crops is a source of revenue leading to high economic sufficiency. While the benefits of crop farming are incredible, the negative environmental impacts are to be considered as well. A bigger population means a high level of food consumption. This high demand prompts farmers to use modern methods of crop production to increase the yield. Crops such as corn that would have previously taken more time to mature for market and consumption, take less time. Taking note of the conservationist's grievances proper land management practices are vital for lessening the already identified impacts on the environment.

In support of this, corrective measures like the use of organic fertilizers instead of chemical fertilizers, for example, farm yard manure collected from livestock and biodegradable materials such as wood, paper, food waste, leaves can alleviate the earlier mentioned issues. These, are natural materials and cannot affect the environment adversely. Similarly, controlled irrigation like the use of drip irrigation is helpful in reducing water wastage, unlike other methods, to thereby reduce soil erosion (Christopher, et al, 2016).

Another preventive measure a farmer can use is practicing agroforestry. This is an agricultural practice that involves growing trees around or among crops or pastureland. This intentional combination of agriculture and forestry has varied benefits, including increased biodiversity and reduced erosion. Controlled livestock grazing or grazing in designated areas only can help in lessening the negative effects caused. For example, a farmer one may decide to graze livestock less often but, provide cultivated fodder instead. Additionally, a farmer may opt for zero grazing against free grazing his or her livestock.

## 3.8 GLOBAL WARMING AND AGRICULTURAL LAND DEGRADATION

In trying to find the causes of agricultural land degradation the preponderance of available literature revolves around the theories of mismanagement of resources and excessive and uncontrolled population growth, especially in various developing countries. A smaller volume of literature argues that the impacts of global warming are the cause of drought and, therefore, of agricultural land deterioration and resource depletion such as water (Chu, et al, 2015).

It has been argued that droughts are a recurring phenomenon attributed to the impacts of global warming. However, the time between consecutive droughts, as well as the duration and intensity of droughts are uncertain. Because of this uncertainty, it has been claimed that planning for the impacts of future droughts is very difficult, especially with regards to agricultural practices.

Researchers have also claimed that anomalies of global warming in different parts of the world can be related to drought episodes. For example, many droughts may have been caused by an atmospheric- oceanic phenomenon called El Nino, which they described as warming of ocean surface water in the eastern equatorial Pacific off the coasts of Peru and Ecuador (Chu, et al, 2015).

Different human activities have also been reported, by various researchers to cause inadvertent modifications in global warming and this has caused drought and deforestation. For example, air pollution in industrialized countries, produced by fossil fuel combustion, has caused acid precipitation which in turn has destroyed trees. The air-borne pollutants travel beyond the countries where such pollution occurs. Again because of industrialization, the level of carbon dioxide in the air is said to have increased, creating a greenhouse effect which has increased the atmospheric temperature and variability in climate.

It is also argued that rainfall in tropical moist forests comes from the loss of water by evapotranspiration and that deforestation reduces this source of rain. Again overgrazing in marginal areas reduces the vegetation cover and, it is reported, that this increases the reflectivity of the earth's surface and decreases what is known as "biogenic nude" that is important in the formation of precipitations and therefore reduces rainfall (Chu, et al, 2015). Climate change has led to drought and human suffering in developing countries. Such hardships are evidenced, for instance, in the scarcity of domestic energy.

## 3.9 POLITICS AND AGRICULTURAL LAND DEGRADATION

Depletion of resources and environmental degradation can be attributed to the commercialization of forests. Trade in forest resources includes timber, poles and wood fuel. While commercialization of forests provides employment, the problem of global warming and resource depletion far outweighs this benefit. The emergence of a commercial market is the recognition that forests have become scarce and, therefore, a valuable commodity. The trade of forests resources does not bring lasting prosperity to the rural areas because the price paid by urban dealers is usually too low. It is sufficient to encourage the rural people to cut free trees but it is rarely enough to encourage farmers to invest in wood fuel by converting some of their lands to growing trees to meet the urban demand. This is how resources are affected by the distribution of power. Therefore, it becomes important in encouraging afforestation. Researchers have argued that the production and distribution of wood fuel as a resource is influenced by urban dwellers. A large number of urban workers consist of rural immigrants who are periodically engaged in wage employment. The growth of these workers' incomes is very low. It can then be concluded from this that in order for the urban wood fuel dealers to stay in business they have to keep the costs low (Weaver, 2011).

It has further been argued that the urban dealers are indifferent to the disappearance of resources. When all the salable wood in an area has been depleted, they simply move to another area. Africa's urban population will double by the year 2020, and it would be the low-income sector that would increase the fastest and, therefore, increase the demand for wood fuel. It has been pointed out that in certain cities, charcoal is the popular domestic fuel, and that because the production of charcoal involves large losses of energy the process of deforestation is thereby accelerated. Charcoal production is particularly destructive because live trees are cut instead of using dead twigs and branches. This will increase the amount of carbon dioxide in the atmosphere hence global warming (Weaver, 2011).

## 3.10 THE WAY FORWARD TOWARDS NATURAL RESOURCE MANAGEMENT

A natural resource is any material that occurs in nature and is beneficial to humans. These include water, air, land, forests, fish and wildlife, topsoil, and minerals. These natural resources can be classified as either renewable or nonrenewable resources. In most cases, renewable resources such as cropland, forests, and water can be replenished over time by natural processes and can be indefinitely sustainable if not overused. Resources that are nonrenewable resources, for example, diamonds, minerals, and oil are found in limited quantities, and their value increases as supplies decreases. A nation's wealth and status in the world economic system is often determined by access to natural resources (Pearson, 2011). Struggle over territory has been the most prevalent form of conflict especially in areas seen to have natural resources that are in high demand and the natural resources often underlie those territorial struggles. Disputes that occur internally always arise due to local environmental degradation, for instance, when the main freshwater source is polluted by factory emissions. When population migration increases, it can cause ethnic clashes because of the demand for scarce resources such as water or timber.

When natural resources are in abundance, then it creates a bigger threat and it can also create conflicts. Some countries like Cambodia and Iraq which have abundant natural resources have experienced what has been coined as the resource curse, economic stagnation and violent conflict over access to revenues. The easily looted resources, which have been made available, such as diamonds, have encouraged rebel movements. Both the Cambodian government and the rebel group Khmer Rouge funded their military activities using revenue gotten from the sale timber. In a country like Iraq, competition over the control of oil production has helped fund the violence among the Kurds, Sunni, and Shiites. As noted, there is a long list of conflicts with a strong link to natural resource exploitation (Pearson, 2011).

Players from outside have used many techniques in conflict and post-conflict situations to shape natural resource management. International demand drives the value of many natural resources, and this has made many outsiders participate in the market for natural resource. Countries that are part of this market, international organizations, foreign investors and trade-partner can easily influence the local market and governance. Cooperation from these key participants is necessary for conflict situations because it can alleviate improper natural resource use that can escalate and sustain conflict. A good example is when outside parties help with negotiations and agreements, suggests solutions such as eco-parks or sustainable development and conservation mechanisms, that are pushed for adherence to international standards, and that they provide the necessary aid and economic incentives to make change reality and also help resolve conflicts. The United Nations passed a resolution and organized arms embargos or sanctions so as to help the natural resource business to be in a better position to curb illicit economies and move toward transparency (Pearson, 2011).

Outside security forces and the UN peacekeepers have managed to establish peaceful zones and barriers around lands that are in dispute, pipelines, oil fields and other natural resource-related structures; that has helped prevent the return of violence. Financial institutions which are recognized internationally such as the World Bank Group have been of great help with advice on revenue transparency and management, economic stability, and financing resource-driven debts. International businesses and their affiliates have been able to can recognize illicit economies and they require certification of legal import, export and trade. Finally, Non-governmental organizations are often present in post-conflict situations so as to deliver humanitarian relief, assist in resettling refugees and persons who are internally displaced (IDPs) who have been forcefully removed from their areas which are rich in resources. These NGOs also advocate for enforcement of international standards and other methods of natural resource wealth management as well as and monitoring human rights violations (Pearson, 2011).

Final note: This chapter described the importance and current depletion of natural resources and the effect of this depletion on agriculture. Solutions to potential challenges were also discussed to tackle the distribution of resources to maintain peace.

# Relationship between Global Warming and Agricultural Crop Farming

# 4

## CONTENTS

4.1 Introduction.......................................................................................76

4.2 Effect of Increased Temperature on Growth and Development of Agricultural Crops ..............................................76

4.3 Impacts of Soil Salinity on Crops Due to Global Warming ................79

4.4 Impacts of Global Warming on Agricultural Soil Properties................82

4.5 The Connection Between Minimum Tillage and Global Warming ......87

4.6 Excessive Use of Agricultural Fertilizer Leads to Global Warming......89

4.7 Transforming Agricultural Practices ....................................................91

## 4.1 INTRODUCTION

There are numerous relationships between the impacts of global warming and agricultural crop farming. Research has shown that global warming has affected agricultural crop farming in one way or another. On the other hand, there are several agricultural practices such as the application of nitrogen fertilizers have contributed to global warming (Stevanović et al, 2016). Therefore, this chapter is going to discuss some of these relationships in a more elaborate way.

## 4.2 EFFECT OF INCREASED TEMPERATURE ON GROWTH AND DEVELOPMENT OF AGRICULTURAL CROPS

The tropical environment in which most agricultural crops are most abundant is characterized by high light intensity, high ambient temperature and intermittent low soil moisture due to the effects of global warming. The radiation from the sun controls directly the ambient temperature and indirectly influences soil moisture through its influence on rainfall and evapotranspiration. Therefore, to evaluate the nature of the water relations of agricultural crops in the tropical environment, it is necessary to know the response of these plants to light and variations of ambient temperature (Ariel and Robert, 2011).

Plant water relations are directly controlled by light through its influence on stomatal movement and indirectly through its influence on leaf and air temperature. Light controls the general growth of a plant directly through its involvement in the assimilation of the carbon dioxide and indirectly by its effects on processes such as endogenous rhythms, flowering, and germination.

Furthermore, light affects photosynthesis both directly and indirectly. It offers the necessary energy quanta which set the complex processes of the light reaction into action. The energy which is absorbed by chlorophyll molecules is used in the splitting of one of the hydrogen bonds in the water molecule through a process known as an oxidation-reduction reaction involving accepting and donating electrons.

Although a large amount of information is available on the influence of light on photosynthesis in general, there is still a lot which can be learned about the subject. Information is still needed about the way the photosynthetic efficiency in agricultural crops is related to their productivity (Ariel and

Robert , 2011). Light and leaf water content are the most important factors regulating stomatal apertures.

**Figure 4.1:** In tropical countries, agricultural plants are usually subjected to severe water stress at high light intensity due to the effect of global warming.

**Source:** https://cropwatch.unl.edu/2016/impacts-extreme-heat-stress-and-increased-soil-temperature-plant-growth-and-development

However, it is not clear why the lack of an apparent light saturation point associated with a plant does not lead to excessive water loss and finally act as a disadvantage on the growth of these plants in the dry tropics. The influence of light on plant growth is closely related to the influence of temperature. Temperature affects plant growth both directly and indirectly. Temperature activates all the metabolic processes and also determines the upper and lower limits of heat energy suitable for proper growth. Such limits are generally called cardinal temperatures and different plant types are characterized by different cardinal temperatures (Ariel & Robert , 2011).

Agricultural plants have different cardinal temperatures from plants. Indirectly, temperature influences plant growth by controlling vital processes such as water movement in the:

- Soil-plant-atmosphere continuum
- Photosynthesis
- Respiration

- Differentiation
- Breaking seed and bud dormancy

In the water relations of plants, temperature controls the vapor pressure gradient between the leaf and air. Also, the air temperature controls the vapor pressure of the evaporation surface in the mesophyll through its effects on leaf water potential. Temperature controls the movement of stomata and generally, stomatal opening follows the temperature curve of carbon dioxide assimilation.

In different plant species, the temperature of brightly illuminated leaves is near air temperature if the latter is about 33 °C. Below this temperature, leaves tend to be warmer than air and above this temperature leaves are cooler than air. This distinctiveness is partly due to the exponential increase in water vapor pressure with temperature, resulting in increased negative feedback of transpiration and partly due to the increase in stomatal aperture with temperature. Leaf temperature is influenced by three following components (Christopher et al, 2016).

- The wind speed
- Relative humidity
- Transpiration

Therefore, by reducing the value of the boundary layer resistance, increasing wind speed tends to increase transpiration losses of water in the low wind speed range. If the leaf temperature is above air temperature due to radiant heating then increasing wind speed causes cooling and reduces the vapor pressure gradient between the leaf and air.

The ability of thermophilic plants to sustain active photosynthesis at 35-50° C is associated with the presence of a heat stable protein synthesizing mechanism. This mechanism is said to enable plants to replace enzymes lost by heat denaturation. The failure of these plants to maintain photosynthesis at temperatures below 10-12° C is suspected to be associated with a breakdown in the synthesis and function of macromolecules concerned with the formation of chlorophyll and enzymes involved in the carboxyl action of carbon dioxide. Temperature does not only influence leaf water potential but it also controls stomatal conductance. It has been revealed that by promoting stomatal opening through the increase of insolation it is possible to increase transpiration (Christopher et al, 2016).

## 4.3 IMPACTS OF SOIL SALINITY ON CROPS DUE TO GLOBAL WARMING

The primary expected effects of global warming are attributed to high temperatures, changes in precipitation patterns (lower and more unpredictable precipitation) and higher recurrence of outrageous events. Global warming and climate change have very adverse impacts on the quality of agricultural soil. One of the main impacts is an increase in soil salinity. This is mainly attributed to the high rate of evaporation of soil moisture due to high temperatures leaving behind high salt content in the soil (Douglas & Charles, 2018).

Salinity is the occurrence of high concentration of dissolved salts in a solution. Therefore salinity in the soil would mean a high concentration of dissolved salts in the soil solution. Determination of the salt status of a soil may be done in either of the two ways.

- The gravimetric method of salt determining
- Calculating the electrical conductivity of a saturated soil extract

Agricultural land whose saturated soil extract has a conductivity of 4 m mho $cm^{-2}$ or more is saline. In most cases, the pH reaction of saline soils ranges from 7 to 8.5. The origin of soil salinity may be contributed to by many factors and events; therefore even though most of the saline soils occur in dry semi-arid areas, pockets of saline soils may also be found in wetter areas. The most prevalent factors known to have caused the problem of salinity in the past include impeded drainage, infiltration by saline water, and distillative precipitation (Douglas & Charles, 2018). This is the reason behind the apprehension that exists about irrigation projects.

However, deposition of salt caused by evaporation of soil moisture it is the most feared cause of soil salinity. This phenomenon is contributed by high temperatures caused by global warming and climate change, especially where drainage is poor and the water scarce or of low quality. Global distribution of saline soils has been reviewed by many researchers. Field evaluation studies in various countries reveal that saline waters and soils occur in various parts of the world. Most of these studies have been carried out, mainly, in areas marked for irrigation projects or some large-scale agro-based ventures. Soil salinity is important as an area of study because it impairs the growth and development of most plants except the halophytes (Douglas & Charles, 2018). Most agricultural crops are glycophytes. They do not resist high salt in the soil. Therefore soil salinity is a threat to food production; by reducing total acreage of arable land. Nevertheless, some of the crop plants have been shown to tolerate low to medium salinity. Thus, there is non-uniformity in the susceptibility of crop plant species and varieties to salt stress.

**Figure 4.2:** Some of the major impacts of soil salinity on crop plants may range from impaired germination to the death of the plants.

**Source:** https://www.wsj.com/articles/SB10001424052702303292204577517181109756756

Therefore, there has been a lot of work carried out to find out the mechanism and basis of the detrimental effects, with a view to isolate of tolerant varieties and discernment/bf markers. Among the agricultural crops that have been intensely studied in relation to impacts of soil salinity include:

- Barley
- Wheat
- Sorghum
- Pearl millet
- Rice
- Maize

Different aspects such as morphology and metabolism have been evaluated under salt stress. Generally, the study on plant response to soil salinity is divided into two major stages;

- *Germination*: These studies are mostly carried out in the laboratory using paper-lined Petri dishes irrigated with the treatment solutions.
- *Growth and development*: These studies are carried out in soil and, occasionally, nutrient solution.

Researchers have revealed that salt stress generally impairs germination and growth and development of the crop plants. Most of a plant's dry matter is accumulated by way of photosynthesis; therefore this process plays a key role in the development and yield of crop plants. This is the reason for which the photosynthetic efficiency, translocation of photosynthates and the composition of the photosynthetic pigments (mainly chlorophyll) in salt-stressed plants attracts great interest.

In cereal crop plants and grasses in general, salt stress has been shown to change chlorophyll composition and the absolute chlorophyll content. This could lead to reduced photosynthetic rates. The reduced photosynthetic rates may result from the above causes; alternatively, it may result from reduced translocation or enzyme activity due to low water content or closure of stomata (Bhupinder et al, 2011).

It is believed that agricultural plants under salt stress may suffer from either over-accumulation of mineral ions in the tissues or an ion imbalance whereby the plants absorb more of the neutral than nutrient ions. The ion excess is believed to impair enzyme activity. Certainly, the accumulation of sodium and chloride ions, sometimes accompanied by a decrement of potassium and calcium ions, has been observed for a variety of cereal crops such as barley, in wheat, and millet. Relevant is the fact that some of the fairly drought tolerant cereal crops (barley, millet, and sorghum) are among the most salt resistant crop plants. However, researchers have pointed out that drought-resistant plants may not necessarily be salt resistant and vice versa. However, former observations make finger millet an interesting crop plant to work with in this aspect (Bhupinder et al 2011).

Soil salinity retards the growth and development of plants by interfering with the metabolic processes. The plants which are under salt stress face two distinct problems.

- *The first problem is related to ion excess in the protoplasm*. The high ion content in the protoplasm may interfere with metabolic activities because of the agglutinating or dispersing activity of the ions on the cytoplasm components. Moreover, excess univalent ions have been shown to aggravate the loss of substances from the cells. There are several ways by which a plant may resist the high ion content in the tissues. The first of those adaptations may be the synthesis of ion-resistant enzymes.

- *The second problem is related to the soil-plant water relations.* The addition of saline water to the substrate lowers the latter's water potential. Depending on the amount of water and the irrigation regime the matrix or the solution osmotic potential may have an upper hand in this process. This condition necessitates that the affected plants lower their water potential in order to continue absorbing water from the substrate. In higher plants, this is achieved by absorbing the mineral ions from the substrate. Most plants differ in their ability to carry out this osmotic adjustment and this may sometimes be reflected in their differences in salt resistance. The more resistant plants adjust faster than the susceptible ones. However, this may not always happen.

Bhupinder et al (2011) have shown that plants (halophyte or glycophyte), when compared with their counterparts, are relatively more efficient in fixing carbon dioxide under salt stress. However, the photosynthetic enzymes have been shown to be very sensitive to soil salinity. This is an indication that in the cell these enzymes are not exposed to high ion concentrations. In summary, a plant under salt stress must carry out osmotic adjustment and at the same time avoid the accumulation of excess ions at the sites of enzyme activity, or alternatively synthesize salt-resistant enzyme systems to survive the soil salinity. It is apparent that most plants are ion excluders. Therefore, the resistance of a plant to salt stress will depend on how the plant carries out osmotic adjustment, and at the same time protect its sensitive enzymes from excess ions. In both cases, the halophytes are better adapted than glycophytes.

## 4.4 IMPACTS OF GLOBAL WARMING ON AGRICULTURAL SOIL PROPERTIES

### 4.4.1 Organic Carbon

Organic matter is a significant source of nutrients for most agricultural plants. Organic matters also help in the maintenance of soil nutrients and acidity status. Therefore, the total amount of organic matter in the soil is a very important element that needs to be considered by farmers. Under natural conditions, the effects of global warming and climate change are the most important factors affecting the soil organic matter content. In tropical Africa for instance, rainfall is mainly found to be greatly correlated with the content of organic carbon in the soil. Generally, sandy soils and soils in low rainfall areas have been found to contain less than 2% of organic matter. On the other hand, soils in areas with high rainfall and high altitudes have been observed to contain over 6% organic matter. In savanna African regions, the organic matter content has been found to range between 0.08 - 4.7% in under well-drained soils and 0.19 - 11.0% in under poorly drained soils (Douglas & Charles, 2018).

There are two major factors that influence the organic matter content in the soil. These are:

- The rate of addition of organic matter in the soil
- The rate loss of organic materials in the soil

In virgin agricultural land, a dynamic equilibrium is established between the gains and losses so that the organic matter content is maintained at a constant level. The following equation is used in relation to the equilibrium conditions of the soil.

$$C = \frac{bm}{k} \; and \qquad a = bm$$

Where;

C represents the percentage of organic carbon in soil under equilibrium conditions. Its unit is given in terms of (tons/ha)

B represents the total annual amount of fresh organic matter that is added to the soil in terms of organic carbon. Its unit is given in terms of (tons/ha)

A represents the total amount of soil organic carbon added in a year. Its unit is given in terms of (tons/ha)

K represents the total decomposition constant of the organic carbon in the soil

M represents the rate of conversion of the fresh organic matter into soil organic carbon. It is expressed in terms of percentage.

Both b and k are the parameters that change when the equilibrium conditions are disturbed. For instance, climate change causes a reduction in b and an increase in c. The rate of decomposition (k) depends greatly on:

- Impacts of global warming on temperature, rain, and relative humidity

- Climate change
- Cultivation treatment
- The cropped vegetation

Adverse climate change frequently leads to a decline in the soil organic matter. This decline in organic matter can generally be attributed to two causes. First, clearing and cultivation of land result in a reduced rate of addition of fresh organic material. Secondly, the rate of decomposition of the soil organic carbon is accelerated as a result of a combination of factors favoring increased mineralization after the matter is highest just after clearing and reduces or stabilizes with time (Douglas & Charles, 2018).

### 4.4.2 Total Nitrogen and Carbon Ratio

A lot of soil nitrogen is bound in the organic matter of the soil. Therefore, as the organic carbon content of the soil declines so does the total nitrogen level. The rate of mineralization of nitrogen is influenced by temperature (which is a key element of global warming) and the mode of cultivation (Bhupinder et al, 2011). The difference in nitrogen content in soils is attributed to a decrease in mineralization of organic nitrogen in soils under direct drilling. This is further ascribed to the presence of favorable conditions to mineralization in the cultivated environment as compared to that under reduced cultivation.

In addition to total nitrogen reduction, climate change also lead to a decline in all forms of nitrogen except non-exchangeable nitrogen. Untilled soils normally have higher nitrogen content near the surface. Agricultural land under reduced tillage result in a significant increase in total nitrogen compared with tilled soils.

It has been shown that the carbon to nitrogen ratio tends to be lower for soils of arid regions as compared to those of humid regions. The addition of organic residues with a wide C/N ratio will result in competition for nitrogen between the soil microorganisms and higher plants. The C/N ratio of soils is therefore important in the nutrition of plants (Bhupinder et al, 2011).

### 4.4.3 Cation Exchange Capacity

There are several components of soil that influence the capacity of soils to adsorb and exchange cations. Some of the most important components are the amount and the type of clay and the organic matter content in the soil.

The organic matter content is, however, the main important component of the soil that is altered by the effects of global warming, climate change and the mode of cultivation. Organic matter is also important in contributing to the cation exchange capacity in soil. Other soil components that affect the cation exchange capacity are the soil pH and the electrolyte content of the soil solution. Cultivated soils under different tillage and fertilizer treatments have shown a decline in cation exchange capacity. The intensity of tillage and the severity of the rotation program are also reported to influence significantly the cation exchange capacity of cultivated soils. Despite the high correlation between the organic matter content and the cation exchange capacity, it should be noted that the changes in organic matter content do not always imply a corresponding change in the cation exchange capacity status of the soil in question (Jurg & Gregory, 2014).

The changes in cation exchange capacity as a result of deteriorating soil properties due to the effects of global warming and climate change are influenced by the percent or severity contribution of a deteriorating component to the total cation exchange capacity of the soil in question. Therefore, a soil with a high percent of cation exchange capacity due to high temperature will show significant declines in cation exchange capacity as the temperature of the soils declines.

### 4.4.4 Soil pH

Soil pH is directly related to soil salinity which is brought about by the effects of global warming and climate change. Most tropical soils, however, do not have acidity problems as most tropical crops are adapted to acid conditions (Jurg & Gregory, 2014). The direct influence of climate change to the soil pH is related to the changes in the soil buffering capacity due to changes in soil organic matter and the application of lime and fertilizers. Most reports in literature on the effects of global warming on soil pH are not consistent as the effects depend on the cultural practices employed in the cultivation. The use of nitrogen fertilizers has been to cause a decline in pH; this is especially so with the ammonium- based nitrogen fertilizers. For instance, the addition of soil amendments such as compost and liming material is noted to increase soil pH under continuous cultivation in silt loam and Podsol, respectively.

### 4.4.5 Bulk Density

The bulk density expresses the ratio of the mass of dried soil to its total (bulk) volume. The bulk density is affected by the structure of the soil and its swelling and shrinking properties. These two components are directly associated with the effects of global warming. Consequently, the measure of the soil bulk density is an important soil parameter that is frequently made to evaluate its structure. The higher the bulk density the more compacted the soil and the lower the pore space. Compaction in soil leads to poor soil physical properties such as low infiltration, poor aeration, and high impedance to root penetration. Bulk density for the surface layer of fine-textured soils generally ranges from 1.0 to 1.6 $g/cm^3$. In sand and sandy loams, the bulk density may run as high as 1.8 $g/cm^3$ because sand particles tend to pack closely and remain in close contact (Jurg & Gregory, 2014).

In reference to the impacts of global warming and climate, the following are some of the key elements that are associated with soil bulk density.

- Organic carbon content in the soil
- The soil particle size composition
- The intensity of cultural operation
- The crops cultivated
- The tillage practice

### 4.4.6 Aggregate Stability

Aggregate stability refers to the resistance of soil to the influence of water and mechanical manipulation. Temperature is one of the main elements of global warming. Researchers have reported the close association between soil aggregate stability and change in soil temperature. Soils with poor aggregate stability result in poor structural stability. Poor structural conditions may cause yield declines through a direct influence on root growth, deteriorated drainage conditions and hence poor aeration and poor nutrition. The aggregate stability is a relative quality and the methods used to characterize it are arbitrary. The most common method involves wet and dry sieving of soil samples under certain arbitrary but well controlled and defined experimental conditions. These conditions, however, subject the soil to forces which are not encountered in the field. Wet and dry sieving does not thus represent the actual stability of aggregates in the soil under field conditions. The wet and dry sieving is however a very invaluable method for rapid determination of the soils relative aggregates stability. The degradation of aggregates is highly correlated with the soil organic matter content. Therefore, a reduction of soil organic matter due to the effects of global warming has a great effect on the stability of soil aggregates (Bhupinder et al, 2011).

### 4.4.7 Soil Erodibility

Soil erodibility is the vulnerability of soil to erosion due to its inherent characteristics. Normally, global warming causes high atmospheric temperature which leads to more vigorous hydrological cycle such as extreme precipitation events. This is more likely to cause soil erosion. It has been claimed that, although soil loss depends on impacts of global warming and climate changes, the inherent soil characteristics are the most important determinants.

Some of the factors that influence erodibility of the soils are:

- Particle size distribution
- The kind of the ions on the cation-exchange complex
- Type of clay mineral
- Organic matter
- Organic matter content and cementing materials in addition to clay

Soils low in organic matter and high in silt and very fine sand frequently form aggregates which are relatively unstable. These aggregates are easily destroyed by the beating action of rain. Raindrop energy and fine grains flowing into and plugging surface pores combine to produce a dense compact layer at the soil surface. As a result, there is reduced infiltration which results in increased runoff and erosion (Jurg & Gregory, 2014).

Soils with high organic matter contents have more stable aggregates because of the strong bonds between their colloids. The cohesion and internal frictional strength determine a soil's shear strength which will influence a soil's susceptibility to detachment. The higher the moisture contents the lower the cohesion and internal frictional strength of soils. Large and stable aggregates make soil difficult to detach and transport. An important point about the size of aggregates is that the larger they are, the greater the delay before they are reduced to fine 10 particles which can be transported, and the higher the surface retention capacity of the soil surface, hence the greater the infiltration (Jurg & Gregory, 2014).

## 4.5 THE CONNECTION BETWEEN MINIMUM TILLAGE AND GLOBAL WARMING

Most farmers in the world have been practicing minimum tillage as a way of preventing soil erosion. When incorporated with other types of agricultural practices such as pesticide and effective nutrient management, minimum tillage can actually help in reducing soil degradation. Minimum tillage or reduced tillage is a tillage system which creates suitable soil conditions for crop seed germination and growth by using a minimum number of operations. An example would be to prepare a seedbed in planting rows only, leaving the area between rows undisturbed. Weed control is facilitated by spraying contact herbicides and soil residual herbicides on the land immediately the following planting or before planting to eliminate existing weeds and to prevent further germination of the weeds and/or slashing or a minimum amount of hoeing. This is in contrast with zero tillage where the crop is planted directly into an undisturbed or unprepared seedbed. The entire previous residue is left on the surface. Weeds are controlled by the use of herbicides, burning or slashing (Manjit & Surinder, 2013).

**Figure 4.3:** Minimum tillage.

**Source:** https://globalfarmernetwork.org/2016/10/advocating-agriculture-changing-planet/

When minimum tillage is practiced in the context of global warming and climate change, most farmers are able to reduce the emission of various greenhouse gases (methane, carbon dioxide, and nitrous oxide) that are attributed to global warming. In other words, minimum tillage helps to reduce the loss of nitrogen and carbon. These are the two major elements associated with impacts of global warming such as an increase in global temperatures, changes in precipitation patterns, and a decrease in ice cover among others. Global warming is attributed to a loss of soil moisture thus, affecting agricultural productivity in one way or the other. This mainly occurs through evaporation and intense surface run-off. Through the practice of minimum tillage, farmers are able to conserve soil moisture.

A lot of work on zero tillage and/or minimum tillage has been done in many countries. Maize and soybeans are the most widely grown crops under these tillage systems with a greater part of the area currently under cultivation being in the United States of America. The use of pre-emergence herbicides requires a clean and well-prepared seedbed. Reduced tillage system decreases time and number of operations required to establish a seedbed. It also reduces soil erosion and moisture loss from the soil and may enable quick and safe use of steep landscape. One major problem associated with reduced tillage is the shifts in weed flora in favor of perennial weeds. However, this may not become a serious problem if proper crop husbandry practices are undertaken (Manjit & Surinder, 2013).

## 4.6 EXCESSIVE USE OF AGRICULTURAL FERTILIZER LEADS TO GLOBAL WARMING

Studies have shown that excessive use of agricultural fertilizers is contributing a lot in terms of global warming through the emission of nitrous oxides. Most of the nitrogen that is not absorbed by the plants normally react with soil and emit dangerous greenhouse gases. Statistically, over 80% nitrous oxide emissions in the whole world come from the use of agricultural fertilizers. This is contributing about 8-15% of all greenhouse gases causing global warming in the world (Ariel & Robert, 2011).

**Figure 4.4:** Excessive use of agricultural fertilizer leads to global warming.

**Source:** https://wbmfoundation.wordpress.com/tag/fertilizer/

The use of fertilizers in agricultural crop production has been on a rise day after day. In African countries, this increased rate of fertilizer consumption can be attributed to the sub-division of formerly European owned estates into smallholder farms, the approval of the government for smallholders to grow cash crops, the introduction of hybrid crops and the introduction of fertilizer subsidy. A combination of rapidly increasing fertilizer prices and insignificantly increasing farm produce prices have led to downward adjustments in optimal fertilizer application at the farm level (Ariel & Robert, 2011). Globally, two-thirds of the available fertilizers are applied to cash crops such as coffee, tea, and sugar cane while one-quarter is applied on maize and wheat and very little on other crops. The largest share of the global total is used by large-holders and estates though application by smallholders has been increasing over time.

Fertilizers exist as solids and liquids. They are utilized to improve plant development and plant wellbeing. Aqueous fertilizers are more efficient than solid ones since they directly provide the plants with the nutrients. They are easily absorbed by the plants and hence are a quicker fix. Not all fertilizers are made out of similar elements. The elements applied often that are nutritious and help improve the health of plants include Nitrogen, Phosphorus, Potassium as well as Magnesium. These mentioned elements help facilitate the process of photosynthesis. Potassium lessens water loss and shriveling in plants while Nitrogen is basic for chlorophyll amalgamation. Magnesium kills causticity in plants and acts are a transporter for phosphorus since it can move around inside the plant tissues effortlessly. Phosphorus converts energy from the sun into the plant and a lack of it leads to discoloration.

Most crops do not always respond to potash as most soils around the world have high enough levels. Nitrogen has been reported to increase crop yields. Nitrogen affects the quality of crops in various ways.

- Applied nitrogen increases the dry matter and protein contents without decreasing the starch content obtained by applying phosphorus and potash.
- In potatoes, increased rates of applied nitrogen decrease tuber specific gravity and increase the number of tubers with light colored skins, excess nitrogen affects flavor and tuber texture.
- The amount of nitrogen required is influenced by previous cropping and rainfall in the proceeding season. This may be because these two factors influence the amount of organic matter added into the soil and thus the nitrogen level in the soil.

- As nitrogen became available in large quantities, yield and grade normally go down, probably because of too lush haulm development at the expense of tuber development.
- Lack of adequate nitrogen results in a small leaf area and thus a slow bulking rate and ultimate low yield.
- Nitrogen is also essential later in the season to prolong the bulking period.
- Need for nitrogen under tropical conditions may be more acute due to losses through leaching.
- Phosphorus has been reported to increase yield in medium and small grades by generally increasing average weight for the produce.
- Excess phosphorus may reduce yield by premature ripening of the crop.
- Potash increases the yield of large tubers and depresses dry matter content but tuberous crops do not always respond to potash in some soils as they already have high levels.
- Excess potash can give rise to magnesium deficiency. It is thought that imbalance between nitrogen, phosphorus, and potassium aggravates after-cooking blackening.

Application of nitrogen, potassium, and phosphorous fertilizer decreases dry matter and starch contents and slightly increases the ascorbic acid contents. Giving more nitrogen and potassium almost halves the amount of phosphorus needed for maximum yield by increasing the rate of phosphorus response. The little work carried out in fertilizer trials by different researchers indicates that only nitrogen and phosphorus are important for good growth in most crops (Ariel & Robert, 2011).

## 4.7 TRANSFORMING AGRICULTURAL PRACTICES

In most of the traditional agricultural practices, there are comparatively inefficiencies in the allocation of factors of production. This has made most of the traditional farmers to remain poor but efficient. This implies that no appreciable increase in agriculture can be had by reallocating the factors at the farmers' disposal since all factors are already fully employed. To increase production, in the current situations attributed to global warming and climate change, there is a need for a total transformation of the agricultural practices.

According to the views given by Christopher et al (2016), the transformation of various traditional agricultural practices into a highly productive sector depends on the investment made on agriculture and form it takes to make it profitable. In most countries around the world, agricultural practices (both livestock rearing and crop production) are treated as a source of economic growth which can act as an engine of development. However, the form of investment is significant for the realization of this goal.

To increase agricultural production, especially in arid and semi-arid areas, the government should introduce incentives to guide and reward farmers. These incentives are seen as an important component of the investment. In most countries, incentives for the farmer to work more are weak because the marginal productivity of labor is very low and incentives to save more than they do are weak because the marginal productivity of capital is also very low. There is little adoption to change and farmers are more secure about what they know about traditional agriculture than adopting and learning how to use new factors of production (Christopher et al, 2016).

The types of risk and uncertainty about yield associated with the advance in knowledge are of real concern to the farmers who produce so little barely enough to meet subsistence requirements. Researchers have suggested that traditiona l agriculture is resistant to change because introducing new factors mean coping with problems of risk and uncertainty associated with the productivity of the new factor. The rate at which traditional farmers adopt new factors is therefore subject to the allowance for risk and uncertainty.

Agricultural revolutions are currently suppressed due to a lack of adequate incentives. In different countries, the state of incentives is such that it is unprofitable for farmers to undertake modernizing investment that would increase the productivity of agriculture, bearing in mind the various challenges been experienced such as climate change and other harsh weather conditions. An incentive, in this case, is the product of economic information from which the farmer derives his expectations (Jurg & Gregory, 2014).

The cost expectation encompasses:

- The rate of interest
- Rent on land
- Payment for equipment, fertilizer, and labor
- The farmer's time devoted to farming

On returns, expectations encompass the value placed on the farm products to be utilized in the farm household and the expected price of the product to be sold. Optimum economic incentives provide information that leads producers to allocate resources in ways that result in maximum production. Governments by various means often alter open competitive market incentives. The technological possibilities have become increasingly more favorable but the economic opportunities that are required for farmers in the low income countries to realize their potential are far from favorable. Due to a lack of profitable incentives, farmers do not make the necessary investments including the purchase of superior inputs. Therefore, government intervention is the primary cause of a lack of optimum incentives. However, it becomes imperative to determine the conditions that are both necessary and sufficient to attain the optimum increase in agricultural productivity. The use of better agricultural inputs and techniques have been seen to be necessary, but the availability of superior seeds, fertilizers, pesticides and animals along with other forms of new capital is not sufficient to achieve large increases in agricultural productivity (Walter et al, 2014).

The allocative role that producer incentives play in attaining the optimum increase in productivity is very important. There has been real progress in scientific agricultural research oriented to the requirements of poor farmers, especially in areas that are adversely affected by impacts of global warming and climate change, however, the utilization of the contribution of this research is being impaired by distortions of producer incentives. Having achieved real progress in agricultural research along with considerable additional capital for agricultural development, the primary constraint that currently accounts for the persistent disequilibrium is lack of optimum incentives available to farmers. In Africa for instance, there has been a lot of technological breakthrough in agriculture with the introduction of improved crop varieties and livestock breeds that perform better in adversely affected areas. Research breakthroughs and innovations in agriculture are therefore not the most pressing constraints hindering increased agricultural production. The issue is what should be done to establish suitable incentives which can accelerate the use of the research findings (Walter et al, 2014).

There are certain traditional husbandry practices with insurance techniques that should be embraced. Such practices include intercropping and staggered planting to allow a variety of crop inventory and flatten out seasonal labor peaks. Where agricultural inputs are purchased, the farmers' willingness to purchase them is limited by his expected yield which is characteristically low for subsistence farmers. The expectations of yield likely to accrue from the use of credit will determine his willingness to incur debts. Farmers' expectations are therefore an important consideration in the impact of programs to increase production. Farmers' priorities, however, can change but at a slower rate, therefore incentives are needed to change them faster. Small-scale farmers are known to be risk averse and this may be a major hindrance to the use of agricultural credit and improved farming practices. This has been shown to be true in various arid and semi-arid areas in Africa (Walter et al, 2014).

There is the need to establish whether risk averseness explains the farming practices in these areas and what can be done about it. Some of the major factors impeding progress in the agricultural sector include the following:

- The importance of land tenure arrangements in peasant agriculture
- Labor difficulties due to the seasonal nature of peasant farm organization
- Unsatisfactory marketing arrangements for farm produce and long distance or poor communication resulting in high transport cost hamper the peasant farmer as these may make the sale of surplus unnecessary and not worthwhile, thus hindering agricultural growth
- Poor farming practices in peasant agriculture.

It is therefore important to know the problems facing peasant agriculture if they are related to raising agricultural productivity.

Final note: This chapter discussed the scenario of agricultural practices in the context of global warming.

# The Impacts of Global Warming on Key Agricultural Crops

# 5

## CONTENTS

5.1 Introduction .......... 96
5.2 Maize Farming .......... 96
5.3 Potato Farming .......... 97
5.4 Tea Production .......... 100
5.5 Sugarcane Farming .......... 102
5.6 Rice Production .......... 103
5.7 Pigeon Pea Farming .......... 105
5.8 Pyrethrum Farming .......... 114

## 5.1 INTRODUCTION

Throughout history, human beings have not been able to master different climate changes brought about by global warming and have remained under the threat of drought resulting in crop failure. In Africa where the major climatic conditions have changed and are not precisely predictable, failures of agricultural crops due to low and erratic rainfall during the crop growing seasons have been frequent. Crop-climate studies in the tropics are essential to improve human understanding of the problem of crop failures through scientifically informed interventions. Numerous agro-climatological studies have helped scientists to better understand crop environment interactions. For instance, it has been understood that the growth of crops in a certain environment is mainly influenced by their ability to integrate the environmental factors, both abiotic and biotic with the influence by humans being the most important.

Rainfall is the most significant climatic factors at work in the plant environment in the tropics. This is because it has the ability to limit the agricultural potential of the land. Some of the major characteristics of rainfall that play this significant role are:

- Rainfall amount
- Rainfall frequency
- Rainfall intensity

These three characteristics are important during the crop growing season. Also, these three characteristics of rainfall have become very unpredictable in most places, especially in tropics. Because of this unpredictability nature, they have made the tropics more risky for agricultural crop production. Let us now discuss the effects on vital crops:

## 5.2 MAIZE FARMING

The task to quantify the relationship between the impact of global warming and crops might have started just before the turn of the last century. However, but the qualitative effect of some of the weather variables, particularly rainfall, seem to have been known from time immemorial. In China, for instance, the measurement of precipitation dates back to the Eastern Han Dynasty (A.D. 25-200). A network of precipitation observations may have come into existence as early as the $14^{th}$ century when the Third Emperor of the Ming Dynasty ordered that reports on rainfall at various places be sent to him. Whereas this was probably done, none of these precipitation records have ever been discovered (Ariel et al, 2012). During the course of research on documents of the China Dynasty kept in the Imperial Palace in Beijing, a set of rain and snow records submitted by governors of different provinces were found. It is likely that such records were used for advisory services on farming. In a dry climate, the harvests are to an extraordinary degree dependent on rainfall. On the other hand, in moist climates crops are diminished by extremely large quantities of rain. While some researchers suggest a negative relationship between seasonal rainfalls and maximum yields occurring during years of below normal rainfall, others suggest a positive relationship of maximum yields being associated with maximum rainfall. These observations are an indication that different models of the rainfall-yield relationship are likely to be appropriate in different climatic regions.

Over the years so much has been done to relate the impact of global warming and climate change to maize yield. Much of the work looked at the influence of weather variables acting jointly or independently on the yield of maize grown at single or various stations using different techniques. It has been noted that the rainfall from flowering to ripeness is the most important and that the rainfall for the ten days following the date of flowering has an almost dominating effect upon the yield of maize. During this period, the yield is proportional to the rainfall amount. If the rainfall is low during this period, high temperature has a very unfavorable effect on yield. It has also been observed that 12.5mm of rainfall are critical; values equal to or greater than this have a larger effect on the development of maize than falls of lesser amounts (Ariel et al, 2012).

The critical nature of moisture about flowering time has also been noted by other authors who demonstrated the relative importance of rainfall amounts during different growth periods of maize. Grain yield reductions greater than 10% as a result of 4-8 days of wilt at silking have been observed. For comparable stress periods, reductions in yield from water deficits during the ear stage have ranged from 21% to 48%. A significant grain reduction (12 to 15%) has been observed after stress during the vegetative period at early ear shoot and ovule development. These observations show that sensitivity to short periods of water stress is greatest at silking followed by the early ear and vegetative stages in order of decreasing vulnerability (Ariel et al, 2012).

## 5.3 POTATO FARMING

Potato is widely grown in areas of the world. Unlike most crops where a "true" seed is used for raising a crop, the potato is mostly propagated through tubers. On an average a tuber has twelve eyes distributed over the surface, the "crow end" having a cluster of bud eyes. Each eye may have more than one bud and from each eye, several sprouts may emerge. From these sprouts, new shoots are produced. Despite the increasing importance of potatoes in the diets of people in the world, potatoes have not always been available to the people at prices most consumers can afford, owing to the large fluctuation in production from season to season. These high fluctuations indicate problems in production. One of the major contributing factors to fluctuations in potato production is impacts of global warming and climate change.

**Figure 5.1:** Potato production has declined due to the impact of global warming.

**Source:** https://foodfirst.org/agroecology-capitalism-and-climate-change/

The best seed growing regions are usually characterized by lower temperatures (often provided by higher altitudes), freedom from certain disease (such as bacterial wilt and blight) and insect pests (such as tuber moth) and by isolation from contaminated areas. Therefore, it should be feasible to establish disease-free areas for seed potato production, especially in tropical countries where the impacts of global warming and climate change are adversely felt. To establish such areas, further research is needed to study the so-called "place-effect" and to determine the probable causes. This "place effect" is sometimes attributed to differences in seed dormancy which can cause retarded or uneven emergence. Other possible influences are tuber-transmitted diseases that differ in their virulence from place to place, variable soil and moisture conditions as well as a possible natural selection of somatic mutations (Ariel et al, 2012). Production of potato seeds is highly recommended in altitudes above 2300 meters. However, most of these areas are characterized by high rainfall which is conducive to diseases such as bacterial wilt and late blight. Altitudes of seed production vary from one country to another.

- In the United States of America, seed area extends along the Northern tier of states adjoining Canada and includes certain areas of high altitude 1000 meters in many Western states; where diseases spread slowly, the soil is productive, volunteer plants are not a problem and the tubers can be stored economically due to the low temperatures prevalent in these.
- In Peru, the seed is mainly produced in the central highlands at 3200-3450 meters above sea level. This implies that seed production is more favored at the intermediate altitudes.
- In Romania, seed potatoes are initially grown at an altitude of 1400 meters and then multiplied at 800 meters above sea level before being grown commercially. This indicates that although potatoes can be successfully grown at low altitudes, it is preferable to raise seed at a higher altitude.
- In Israel, seed potatoes are grown in the Golan Heights at an altitude of 950 meters above sea level. The growth is rapid due to the length of the day and age of the material planted. This is particularly desired to give a quicker maturity period in the shorter growing season of spring.
- In Britain, the greatest advantage is taken of the relatively cool and windy conditions of Northern Ireland, Scotland and Northern England, to produce seeds with minimal virus infection.

One of the major components associated with the impacts of global warming and climate change is increasing temperature. Therefore, increasing altitude in general moderates the average temperature during the growing season so that seed grown at high altitude should be physiologically younger than that grown at a lower altitude. In many areas, the yield and quality of potatoes are kept below maximum by the prevalence of high temperatures at the stage of tuber initiation. Higher temperatures, however, are desirable at emergence and early development up to tuber initiation, when low-temperature results in slow emergence and growth. During the rapid tuber enlargement stage and beginning of maturation also high temperatures increase yield. Low temperatures are desirable during early storage when an earlier break of dormancy is required. This effect of temperature refers to potatoes in general, but it is bound to affect the quality of the seed tuber through its influence on tuber growth (Hakeem, 2015).

Influence of soil temperature on the growth of tubers is greatest when the air temperature is not conducive to tuber growth. Studies have shown that potatoes give maximum yields at a soil temperature of 15-18° C. A lower soil temperature slows the growth of the epigenous part of the plant and decreases starch content of the tubers, and below 4.4° C growth ceases. The contents of phosphorus and nitrate have been found to be much lower at 10-12° C than at 15-20° C. One of the principal causes of nutrient decrease of potatoes in lower soil temperature is the reduced mineral uptake.

Another component attributed to global warming and climate change that interfere or influence potato production if the amount of rainfall. The optimum rainfall distribution for an early variety is a monthly mean of 150 mm. The distribution of rainfall during the growth period is more decisive for the final crop yield than the amount of rain. According to Hakeem (2015), a well-distributed rainfall of 220 mm during the growth period is better than 350 mm in only two months at an altitude of 1500 meters above sea level. A minimum monthly rainfall of 80 mm is necessary for areas above 1500 meters if a potato crop is to be economical.

## 5.4 TEA PRODUCTION

This is one of the most significant cash crops in Africa. Tea production practices are carried out in areas contrasting significantly in their physiographic and climatic conditions. Tea production varies from large commercial estates managed by commercial companies to small production units owned by smallholder farmers. The availability of soil moisture supply is the most important climatic factor governing the overall tea production. Because of the effects of global warming, seasons have changed in different parts of the world.

**Figure 5.2:** High maximum temperatures in excess of 29.5°C have damaged most of the tea plantations in most areas.

**Source:** https://www.beveragedaily.com/Article/2018/05/30/Tea-production-rises-but-FAO-warns-of-climate-change-threat

The effects of solar radiation and temperature cannot be easily separated from the associated factors of humidity and water supply. For instance, high solar radiation because of global warming is characterized by depressed tea production. Research has also shown that a decrease in soil moisture due to lack of adequate rainfall affects the growth rates of tea plants. The minimum annual rainfall necessary for the successful cultivation of tea is 1150-1400 mm. However, a monthly rainfall of less than 50 mm persisting for several months due to climate change affects tea production severely. Such threshold rainfall values are, however, meaningless without a statement of distribution of rainfall particularly in relation to the evaporative demand of the air. The variations in yield appear to be caused by changes in leaf size and shoot length, rather than the number of shoots. Except during drought, the numbers of flush shoots harvested are independent of climatic conditions. An average monthly temperature less than 18 °C and greater than 30 °C is unfavorable for tea production (Ariel et al, 2012).

## 5.5 SUGARCANE FARMING

Despite its potentialities, the Sugar Belt, in India constitutes an area where floods and droughts have become almost annual phenomena over the past few years. Since the intensity of these environmental constraints varies from year to year and from place to place, there is a likelihood of a wide variation in the output of sugarcane per unit area. Natural and climatic factors are regarded as a crucial element in the decision of the producers.

Soil salinity is another effect of global warming and climate change that influence the production of sugar cane in different parts of the world. Research has shown that sugarcane requires soils that are reasonably well-drained and do not contain the excessive salinity of high clay fractions. Clays, loams, marls and calcareous soils are more or less suitable for cane cultivation, while rich, porous clays and alluvial soils on lowland are the most favorable. Another climatic condition and a limiting factor in sugar cane production is rainfall that is also subject to considerable variation. For optimum sugar cane production, the rainfall should be 1250 mm or more per annum, well-distributed; short periods of a few weeks without rain is unobjectionable, provided the capacity of the soil for retaining and releasing water is satisfactory, but the climate should be warm to hot. The unevenly distributed and inadequate rainfall experienced throughout the area not only adversely affect the establishment of new crops but also greatly interfere with the growth and yield performance of plant crops and ratoons alike (Rakesh & Sengar, 2014).

**Figure 5.3:** It seems that the frequent moisture stress experienced in the Sugar Belt may largely account for the poor and stunted growth in sugarcane.

**Source:** https://www.technologytimes.pk/impact-of-climate-change-on-sugar-cane-diseases/

Sugarcane is a tropical crop, demanding for its best performance a high level of soil fertility, a relatively warm temperature throughout the year, and large amounts of moisture and sunshine. For good yields, temperatures should rarely fall much below 21 °C., and there should be a rainfall of more than 1000 mm. spread over at least 9 months in the year unless irrigation is employed. These conditions are partly satisfied over an extensive part of the Sugar Belt. Without irrigation, sugarcane requires approximately 1500 mm of precipitation evenly distributed throughout the growing period (Rakesh & Sengar, 2014).

## 5.6 RICE PRODUCTION

Crop-environment inter-relationships are very complex since the environment incorporates the biotic and abiotic factors which by themselves are inter-related. Similarly, the crop characteristics in terms of growth and development are also inter-dependent. Hence the complexity of the plant-environment system cannot be under-estimated. When studying the interrelationship between a crop and an environmental condition, there is the need for a dynamic model that should indicate the importance or contribution of each environmental factor to the growth and performance of the specific crop in the defined environment.

### 5.6.1 Key Concepts used in Rice Production

*Paddy*: This is the rice crop grown under submerged conditions.

*Paddy fields*: These are the flooded grounds where the paddy is grown.

*Evaporation:* This is the change of state of water from liquid or solid to vapor. It occurs if there is an input of energy (solar radiation) at the moist surface. There has to be a negative vertical gradient of vapor pressure above the surface for this process to occur. The amount of evaporation from a surface depends on the available moisture within the soil and on the rate of diffusion of water vapor away from the surface.

*Evapotranspiration*: This is the total water loss by evaporation from the ground, wet surfaces of the vegetation and transpiration from the vegetal cover. It is governed by factors, both of the surface and climate. Surface factors comprise the availability of water, the type of soil and the nature of the plant covering. Climatic influences, on the other hand, are expressed in the heat or energy available to affect the necessary evaporation of moisture and the ability of the atmosphere to accept the moisture being evaporated. If surface conditions are not restrictive in the sense that water is plentiful and can move readily through the soil so that plants can transpire freely, potential evapotranspiration will be achieved. Thus, potential evapotranspiration is the evaporation that would occur from a large (one or more hectares), wetted and vegetated land surface. The potential evapotranspiration differs from "actual evapotranspiration" which takes place under limited water supply. For paddy, evapotranspiration occurs at the potential rate since the soil is saturated for most part of its growing season (Ouda, 2015).

Despite the fact that rice has wide adaptability to different environmental conditions; different climatic conditions attributed to global warming have severely affected the overall rice production in the world.

Some of the requirements for rice production in general are:

- Damp warm climates
- High temperatures and humidity
- Prolonged sunshine
- Adequate water supply

Variations in these conditions especially water supply and temperatures affect rice production significantly.

**Figure 5.4:** Drought conditions threaten rice farming in most places around the world.

**Source:** https://www.bbc.com/news/science-environment-10918591

Rice reacts differently to various climatic parameters at various phases of its growth. For instance, the ripening phase of rice is the most susceptible to excessive rainfall while a deviation of ±1 $^{0}$C from the mean minimum daily temperature (18 °C) during this phase lowers the rice yields.

The variability of water requirement by rice in different parts of the world is associated with different rice varieties, soil moisture retentions, climatic environments and water management in the rice fields. The following are some of the seasonal water requirement of rice in a few parts of the world.

- In Japan, varies from 685 to 1300 mm
- In Thailand, rice requires 1240 mm throughout the entire season
- In Laos, rice water requirement is 1390 mm.
- In California, rice water requirement is 890 mm.

Studies on water use by paddy rice in different parts of Africa have probably been overlooked since, in the areas where the crop is grown, irrigation water is readily available. This has posed a problem of choice for suitable water management practices for the rice in the continent. Due to this, agronomists tend to assume that water is not a limiting factor for rice agriculture. Despite this, it should be realized that; although water is readily available, there are some aspects of water management in the paddy fields such as scheduling of irrigation to avoid over flooding and leaching of nutrients which are very important (Reynolds, 2010).

## 5.7 PIGEON PEA FARMING

A changing weather pattern in recent years, especially the failure of the rains to come on time has led to serious droughts in several countries around the world. The unpredictable environmental conditions due to impacts of global warming which lead to crop failure may make the farmers reluctant to use high inputs. However, pigeon pea is reputed to do well under the unfavorable weather conditions of the marginal areas. This outstanding ability is attributed to its deep rooting nature which confers on it some level of drought tolerance to survive protracted water stress.

**Figure 5.5:** Pigeon peas that dried up before they could be harvested.

**Source:** https://indiaclimatedialogue.net/2016/05/03/hunger-stalks-scorched-bundelkhand/1-1-a-pigeonpea-crop-that-died-in-the-fields-before-it-could-be-harvested/

The origin of pigeon pea is presently established to be India. There are several ways in which pigeon pea is utilized. It is a major source of protein in vegetarian diets. In India, the dried grain is utilized in a split form referred to as "dhal". In parts of Africa the dried grains are consumed whole by boiling in a mixture with maize and in the West Indies the green mature seeds are usually canned or frozen (Reynolds, 2010).

Matching crop phenology with the environment is advantageous in situations of environmental stresses. The sustained scientific studies on production, management, and utilization of pigeon pea had been limited in the past. Recently, there is wide recognition of the substantial productivity improvements that have been achieved within the cereals, but the grain legumes remain the "slow runners". The yield of pigeon pea in India is estimated at 690 kg per hectare. The figures quoted for pigeon pea in most cases do not indicate whether it was for short, medium or long duration types. The yields of these three groups differ remarkably, especially those of the short duration versus the long duration cultivars. Environmental factors influence the growth and development of pigeon pea to a great extent. The effects of temperature are mediated through fundamental influences on water movement, uptake, and transport of mineral ions, transpiration and translocation and ultimately cytoplasmic function. The adaptation of traditional cultivars of a crop is usually linked to differential responses to photoperiod which in turn are associated with seasonality and latitude. Water stress affects all aspects of plant growth. For instance, pulses are most sensitive to water stress during the late flowering and early pod-filling stages. The effect of water stress may be modified by humidity (Reynolds, 2010).

The cultivation of pigeon pea is spreading to coastal areas in most countries. The humidity in these areas is higher than further inland where the crop is presently cultivated. The environmental factors are influenced by the sowing date ,and plant phenology while dry matter partitioning can be greatly altered by planting date. Flowering at the lower altitude is delayed. The number of fruit bearing branches and the length of the stem over which inflorescences are produced are clearly related to fruit number per plant and are affected by crop density, time of sowing and climatic factors. A large number of pigeonpea flowers abort. It has been suggested that pigeon peas develop only sufficient fruits that individual plants are capable of filling completely (Reynolds, 2010).

This intrinsic "self-regulating" mechanism is yet to be clearly understood and maybe environmental stresses contribute to it. There is a need to understand the effects of environmental stresses on pod-set in pigeon pea. This will lead to (i) choice of the most suitable growing season; (ii) choice of the most suitable genotype; and (iii) more sufficient use of agronomic resources available to the farmer in the marginal areas. For any given location there exists an optimum planting date. This provides the crop with the optimum conditions for growth and development so that reasonable yields can be realized. The date of planting therefore combines the effects of the various environmental factors and as such is an important aspect to consider in understanding the adaptation of new genotypes to a given location.

### 5.7.1 Effects of Water Stress

Basically, water plays three roles in plant development. These are: (i) to mediate the response of the plant to environmental factors; (ii) to provide a mechanism for the integration of plant development; and (iii) to integrate growth and metabolic activity at the cellular level. Water stress can be defined as a phenomenon brought about by a shortage of water disturbing the normal function of the plant. The rainfall pattern in the semi-arid tropics is characterized by alternate spells of drought and abundant water.

These characteristics are attributed to the effects of global warming been experienced in the whole world. Because of these characteristics, crops are therefore subjected to water stress to varying extents and at various stages of their life cycles. Pigeon pea production is mainly concentrated in the transition zone that constitutes the semi-arid and arid lands. The rainfall pattern is bimodal. The crop is sown in either rainy season and usually harvested after six or ten months depending on the maturity group. Hence water stress is realized in the intervening months that have no rainfall.

Both metabolic and physiological processes in the plant are impaired by water stress. Crop plants respond to water stress in diverse ways which vary from species to species and are modified by the environment. Plant response mechanisms should be classified as follows:

- Drought escape
- Dehydration tolerance
- Dehydration avoidance

The survival of plants in water-limited environments may depend on their ability to combine this response mechanism. There are two types of plant response:

- Leaf shedding
- Leaf angle changes

Some of the physiological mechanism responses due to water stress include the following.

- Leaf cuticular wax
- Osmotic adjustment
- Leaf enlargement
- Stomatal behavior
- Photosynthesis
- Translocation
- Proline accumulation

Nearly all metabolic processes in the plant are affected by water stress if the stress is severe and of long duration.

The effect of water deficits in crop production is mediated through its impact on the duration of the crop cycle and canopy development. The economic yield of a crop is dependent on the total amount of solar radiation intercepted and its conversion into the dry matter as well as the efficiency of partitioning biological yield to economic yield. Therefore, a shortened crop duration or limited canopy development will affect economic yield. In pigeon peas several forms of developmental plasticity have been identified. Paraheliotropic leaf movement minimizes the amount of solar radiation absorbed is well developed in pigeon peas and other legumes. Osmotic adjustment is a fairly useful mechanism under water-limited conditions. Studies show that its contribution to economic yield in pigeon pea is not substantial (Ariel & Robert, 2011).

- For plants undergoing progressive stress from seedling stage until maturity, the reduction in light interception is more important than the efficiency of conversion of photosynthetically active radiation to above ground dry matter. An added advantage of the pigeon peas and legumes is their ability to fix atmospheric nitrogen and therefore improve the soil for the succeeding crop. Water stress has been shown to greatly affect the nodulation process.
- Water stress restricts growth and hastens maturity. Severe stress delays the development indefinitely. Relieving the stress leads to a new flush of flowers. Therefore, pigeon peas exhibit developmental plasticity.
- Flowers and fruit abortion are also observed. The critical period in terms of the effect on yield is at anthesis. Water stress occurring during the vegetative phase does not make any difference between the yield of stressed and non-stressed treatments. However, differences in yield are observed between these treatments when stress occurs during flowering. Therefore, pigeon pea genotypes for water-limited environments should be sufficiently early maturing to escape severe drought stress.
- Reduction in plant height is a result of short periods of water stress. Brief periods of drought stress during vegetative or reproductive growth significantly reduces the rate of shoot growth. It has been suggested that mild stresses at the beginning of the vegetative stage may condition the plant to cope with stress better than when these occur in the later stages. It has been observed that plants irrigated throughout the vegetative phase are susceptible to drought when canopy size increases as compared to those subjected to stress earlier.
- When the plants are stressed, leaf flagging, stomatal closure, shedding of leaves, flowers and young fruits occur. Reduced

biomass also occurs when stress is imposed at the early flowering stage as compared to applying the stress at the pod formation stage.

- Water stress effect is more pronounced when it occurs at a time of complete canopy closure, fruit filling, and formation. Premature abscission of flowers and fruits occurs when lupins undergo water stress.

Pigeon pea is reputed to be drought tolerant due to its deep rooting system. Probably it is for this reason that little attention has been paid to the water requirements of this crop. However, recent studies are a pointer towards an emerging need to understand better the physiological basis of pigeon pea response to water deficits. Due to the nature of the stress that pigeon peas are subjected to in the semi-arid areas, it has been suggested that an appropriate plant type for this situation should characteristically be a perennial with deep roots and osmotic adjustment. The utilization of soil moisture is stated to improve if pigeon pea is planted on ridges (Yadav et al, 2011) .

**Figure 5.6:** New pigeon pea cultivars have been developed to curb the effect of global warming.

**Source:** https://www.icrisat.org/super-early-pigeonpea-dodging-climate-change-in-the-drylands/

## 5.7.2 Effects of Temperature

Temperature is among the most important environmental factors affecting plant growth. It influences almost all growth processes and often determines where and when each plant is most successfully grown. Some of the plant processes that are influenced by temperature are:

- Water movement and the uptake and transport of mineral ions
- Photosynthesis
- Transpiration and ultimately cytoplasmic function

Reducing the temperature to 5 °C from the optimum of 15 °C reduces photosynthesis to a greater percentage in pigeon peas. The effect on translocation is mainly on the rate at which the available translocate is transferred to the sink and the removal of the translocate from the conducting elements of the sink. The effect of temperature is realized quite early in the life of the plant. The maturity of seed and its subsequent ability to germinate is affected by high temperature during seed filling. The average seed weight reduces while the percentage of small seeds increases with high temperature.

High night temperature induces flower abscission. At 22 °C, development of pigeon pea appears normal, although all the flowers abscise within 8 hours after anthesis. This is attributed to male sterility resulting from abnormal pollen development and anther indehiscence. This failure of anthers to dehisce has also been observed in soybean. The duration of pod fill and mean dry weights of pigeon pea are reduced due to warm air temperature. Even under conditions of frequent irrigation seed yields and pod number per unit area is negatively correlated with hot weather during the month of flowering. High day temperature is associated with flower abscission. The combination of high soil temperature and large variation in day/ night soil temperature during flowering inhibits reproduction (Chu et al, 2015).

For pigeon peas, the reproductive stage is the one most affected by warm night temperature and therefore a reduction in yield can be solely attributed to flower abscission. In pigeon pea, like most other legumes, flowers are shed to a large extent. Studies on the effect of temperature on pigeon pea have mainly centered on early growth and development. The events before flowering such as the development of sufficient canopy no doubt have an important bearing on the final yield, but a better understanding of the effects of temperature during flowering and pod formation is of particular importance in the semi-arid tropics. The pigeon pea plants usually start their reproductive development after a slow early growth, which at times coincides with unfavorable weather conditions. Earlier studies suggest that the overall growth rate of pigeon pea is influenced by temperature, with slower rates for winter crops.

### 5.7.3 Effects of Relative Humidity

Humidity has a role to play in plant growth and development. The fluctuation in humidity between the dry and wet periods in the semi-arid tropics is likely to influence the growth and development of crops. The literature on this factor in relation to the growth and development of grain legumes is scarce. Investigations have been diverse and have ranged from crops to weeds.

An outstanding effect of humidity on the plant is its influence on transpiration and other related processes, such as stomatal opening, leaf temperature, translocation of nutrients, photosynthesis, and water potential. The rate of transpiration is determined by the vapor pressure deficit between the moist leaf surface and the air. Water loss from the leaf is increased by an increase in vapor pressure deficit resulting from a decrease in humidity. At low vapor pressure deficits, the water loss per unit of the leaf surface is proportional to changes in humidity. At high vapor pressure deficits, the closing of stomata alters this relationship as the transpiration process decreases.

Stomatal opening is an important plant process. Several factors are responsible for the regulation of opening and closing of stomata. The plant exerts control over increased evaporative demand through a reduction in the stomatal opening to maintain a water balance (Shyam et al, 2011). Stomatal closure comes about as a result of large differences in vapor pressure deficits between the guard cells and the air. The process of photosynthesis is also affected by humidity. The differences in dry matter accumulation at different relative humidity are a result of differences in turgor which then alters leaf area of the plants. Some of the plant nutrients move in the transpiration stream. It has been observed that calcium deficiencies in fruit can occur at high humidity due to reduced transpiration (Christopher, et al, 2016).

Leaf temperature is determined by the effectiveness of transpirational cooling. This is because there is a correlation between leaf temperature and atmospheric temperature in most plants. The response of crop plants to humidity is unclear. It appears that the effects of either low or high humidity differ from crop to crop.

Reduction in vegetative growth would lead to a reduced leaf area thereby limiting the availability of photosynthate. Some recent studies have indicated that humidity affects plant growth and development. It has been reported that dry air causes a significant reduction in whole plant biomass. This reduction is attributed to a reduction in transpiration rate and consequent increase in the water potential of the growing cells. It has also been suggested that increased humidity counteracted the inhibitory effects of light. Lateral buds at the basal nodes develop due to an increase in humidity in pigeon peas. Shoot-induced inhibition of the development of root buds is released when humidity increases. Humidity could also have an effect on the reproductive stage of pigeon peas. Similar observations have been made for soybean. Therefore, it is important to understand the effects of humidity during the reproductive stage in grain legumes (Christopher et al, 2016).

### 5.7.4 Effects of Photoperiodism

The effects of global warming influence photoperiodism in different agricultural crops. Day-length varies with seasons and latitude. The grain legumes, like many other plant families, display specific responses to day-length. Depending on the photoperiod requirement for flowering, plants have been divided into three basic categories.

- Short-day plants: These are plants which require short days before they come into flowering
- Long-day: These are plants which require long days
- Day-neutrals (photoperiod-insensitive): These are those plants that do not fall into either category such as flowers regardless of the photoperiod.

Short- day plants can often flower in relatively long days and vice versa for long- day plants. There is another group referred to as short-long-day plants which flower when short days are followed by long days. Photoperiod is mainly considered for its influence on flowering but it may also influence vegetative growth (Christopher et al, 2016).

The mechanism of plant perception of photoperiod is still a subject of much discussion, though it is generally agreed that phytochrome plays a role. The earlier view had been that the phytochrome pigment exists in two convertible forms. It was held that this conversion process determined whether a plant was short-day or long-day. Presently, change in spectral quality may not be the signal to identify day and night but rather the value of irradiance could be the signal. The various diverse influences of Photoperiodism and plant responses are well reviewed. The initiation of phasic changes in grain legumes is influenced by the photoperiod. As a group, grain legumes are quite sensitive to photo-thermal conditions so that the duration of the vegetative phase varies depending on location and date of planting.

Summer grain legumes are all quantitative short-day species. It has been suggested that grain legumes which give a short-day response are of tropical or sub-tropical origin while those responding to long day-length are of temperate origin. It has been revealed that shoot dry weight and leaf area decrease while pod weight increase with increasing number of consecutive short days. Therefore, late planting of a full season cultivar instead of a short season cultivar gives better yield. Under the circumstances of late planting, the full season cultivar will still develop sufficient source size to maximize yield as compared to a short season cultivar (Christopher et al, 2016).

Photoperiods before and after flowering is associated with:

- Reduced branch number
- Total branch length
- Branch node number

Long photoperiod leads to an increased number of flowers which eventually results in a higher number of pods per plant. It has been reported that long photoperiods accompanied by high-temperature cause flower and pod-shedding. The yields of photoperiod sensitive cultivars pigeon peas are more than those of photoperiod-insensitive cultivars under warm tropical condition. The explanation for this is that sensitive cultivars tend to be sufficiently large when flowers appear and as such are able to produce higher yields. It has been suggested that if vegetative growth is completed before plants attain the size for maximum photosynthesis, grain yield is reduced.

Pigeon pea is a short day plant although recently day-neutral genotypes have been found. Pigeon pea cultivars differ widely in time from planting to flowering at different latitudes and within one cultivar at different altitudes. Several studies indicate that longer duration cultivars require shorter day-lengths than earlier cultivars. The studies on the rate of development of pigeon pea as affected by temperature and day-length show that later-maturing types are more sensitive to day-length. Modification in plant morphology resulting from manipulation of time of planting has been in pigeon pea. Practical application in chickpea has been reported (Shyam et al, 2011).

## 5.8 PYRETHRUM FARMING

The most economically important factor in pyrethrum production is the quality of pyrethrins harvested per unit area. The pyrethrins yield is determined by four yield component:

- The number of flowers in the plant
- The flower size
- The dry matter content of the flowers
- The pyrethrins content of the dry flowers

Flower size, dry matter content, pyrethrins content and resultant amount of pyrethrins per flower head vary strongly with rainfall. During the rainy spells, the flowers become bigger, the dry matter content lower and the pyrethrins content and pyrethrins per flower head higher. During the dry spell, the reverse occurs. Most clones show a similar pattern, although the amplitude may differ slightly among clones. The effect of altitude on flower yield is very difficult to assess. Other factors like soil type, planting time, rainfall, care of the crop, introduce such a variation that differences in yield due to altitude may be covered. Fresh flower yield is dependent upon the flower size (fresh weight of 100 flowers), the number of flowers per stem and number of stems produced over the growing period.

In a pyrethrum breeding program, the ultimate objective is to combine an attractive level of expression for each of these traits into one strain which would result in high pyrethrins production per unit area. The number and size of the flowers are negatively correlated as are flower yield and pyrethrins content. Flower size is negatively correlated with dry matter content and flower yield, while the pyrethrins content does not have any relationship with either flower yield or flower size. Pyrethrins yield is positively correlated to flower yield, pyrethrins content, the number of flowers per plant and 100 flowers dry weight. It has also been reported that flower yield is more dependent on the number of flowers per plant than on 100 flowers dry weight. Pyrethrins content is not significantly correlated with any of the other traits. (Ariel et al, 2012).

Final note: This chapter discussed the breeding of economically important crops in the context of global warming to enable a reader to understand the scenario across the world and efficiently be part of successful agricultural practices.

# Global Warming and Livestock Farming

# 6

**CONTENTS**

6.1 Introduction .......... 118
6.2 The Impacts of Global Warming on Dairy Farming .......... 118
6.3 The Effects of Global Warming and Climate Change on Beef Farming .......... 123
6.4 Goats Farming .......... 131
6.5 Poultry Farming .......... 134
6.6 Can Genetically Modified Animals Help In Alleviating The Effects of Global Warming? .......... 137

## 6.1 INTRODUCTION

Just like in the case of agricultural crop farming practices, there are numerous relationships between the impacts of global warming and livestock farming. In this chapter, some of the impacts on main livestock sectors that will be discussed include dairy farming, beef farming, goat farming, and poultry farming. Each one of these sectors is affected by the impacts of global warming in one way or the other. Subsequently, there are several ways on how livestock farming contribute to global warming.

## 6.2 THE IMPACTS OF GLOBAL WARMING ON DAIRY FARMING

There are many places in the world suitable for the production of livestock. However, dairy farming is considerably below the needed standards. It is theorized in this chapter that the effects of global warming produce physiological stress which decreases milk production. Therefore, any way of increasing milk production by reducing physiological stress produced by the effects of global warming and pastures would make a significant contribution to the economy and welfare of most countries in the world. Most dairy breeds are only able to acclimatize where the mean annual temperature does not exceed 65°-70°F (18.3°-21.1°C). Most local farmers, especially in tropical countries, are trying to improve milk production by upgrading the local indigenous breeds. With the increasing demand for milk, the local indigenous breeds would never be able to supply the needs (Manjit & Surinder, 2013).

Global warming changes as discussed are associated with higher temperatures that can be seen in several ways: the price paid by agriculture for global warming has been pitched at an annual range of 1 billion and 2 billion by the middle of this century. Alterations in climate are associated with extreme events and the inability to allow livestock access to such resources during these events. Another challenge is increased price of dairy products due to lowered supply (Hennessy, 2017).

### 6.2.1 Heat Stress on Dairy Cattle

The physiology of dairy cattle may be affected by the environment both directly and indirectly. The most significant direct effects are:

- Air temperature
- Solar radiation temperature

- Humidity (relative and absolute)
- Air-flow

In the tropics, the most acute problem for the dairy cattle is to keep their body temperature from rising beyond physiological limits. This body temperature can be summarized using the following equation.

$$M - E \pm F \pm Cd \pm Cv \pm R = O$$

Where,

M represents the metabolic heat production. Normally, this metabolic heat production will be influenced by the following conditions.

- Level of food intake
- The type of food
- The catabolic or anabolic processes for which the nutrients are used

E represents the heat loss from the skin and respiratory passage by evaporation

F represents the heat lost or gained bringing ingested food or water to body temperature

Cd represents the heat lost or gained by direct contacts between the skin and surrounding surfaces

Cv represents the heat lost or gained by convection

R represents the heat lost or gained by radiation

To counterbalance the effects of rising body temperature, the animal can:

- Seek shade
- Relax
- Enhance vasodilation
- Avoid unnecessary exercise
- Decrease feed intake
- Decrease hair thickness and length
- Increase respiration rate
- Increase vaporization from the respiratory tract and the skin by a combination of one or more of these

The effect of high environmental temperature due to global warming on actual metabolic heat production depends on the way the heat stress is applied (Manjit & Surinder, 2013). If it is acute stress, metabolic heat production rises; if it is a chronic effect, the heat production over a period of time will tend to drop. These thermoregulatory mechanisms, particularly the effect of lowering feed intake and lower basal metabolism, have the effect of lowering milk production.

**Figure 6.1:** The impacts of global warming affect dairy farming differently.

**Source:** https://www.independent.co.uk/environment/climate-change-cows-dairy-industry-affect-a8512266.html

Temperatures in the range of 40°-70°F (2°-21°C) have no influence on milk yield, but above 70°F milk yield decreases, slowly at first with a sudden drop occurring at 80°F. This fall is mostly attributed to a decrease in the feed intake. Apart from this decrease in milk production, the reproductive ability is also impaired. For instance, research studies have revealed that high temperatures have the ability to lower the rate of fertilization in dairy cows, however, most dairy cows that havc bccn studied normally tend to show only short-term effects, rather than the long term effects, which may tend to be different and to which the animals are subjected in most places around the world.

To evaluate the effects of a particular method of alleviating heat stress on dairy cattle, the degree of heat stress should be determined. The role played by the blood and its circulation in an heat-stressed animal is very significant.

It has been shown that hematological changes may indicate a mobilization of defense and establishment of new equilibrium, as well as a breakdown of body functions and deterioration. In other words, hemoglobin levels can be used as an indicator of heat tolerance. For instance, studies have shown that sudden falls in temperature cause a rise in the leucocyte count of the blood of cattle from temperate regions, possibly due to new types of infection attacking the body, but no work as of now appears to have been carried out on the effects of tropical environments on the leucocyte picture (Manjit & Surinder, 2013).

Therefore, if increased heat loads imposed on an animal of temperate origin by a tropical environment cause a stress reaction, this should be apparent from a study of the blood profile. The sum of all non-specific systemic reactions of the body which ensue upon long-continued exposure to systemic stress can be described as the General Adaption Syndrome (G-A-S). This is composed of three stages

- The alarm-reaction
- The stage of resistance
- The stage of exhaustion, which occurs when adaption to prolonged over-exposure to a stimulus has been developed, but cannot be maintained

The temperature has a vital influence on the course of the G-A-S and can itself act as potent systemic stress. The alarm-reaction of the G-A-S is characterized in the hematological responses of neutrophils, lymphopenia and eosinopenia. If the stress is very severe, erythrocytosis would ensue, with the hematocrit and erythrocyte count rising above normal. The neutrophils aid in defense against the stress through phagocytosis of invading disease organisms, while the lymphopenia and eosinopenia result from the increased use of these cells, since they contain substances useful for humoral defense. The erythrocytosis may represent an excessive defense measure to furnish adequate hemoglobin for the increased metabolic needs of an emergency. The hematologic response that results from a stress reaction will depend largely upon the proportion of adrenergic and glucocorticoid hormones discharged, as well as upon the responsiveness of the whole blood-cell system.

Both the production of adrenal hormones and the responsiveness of the peripheral target organs (circulating or stored blood cells, hemopoietic and blood-destroying tissues) depend very considerably upon conditioning factors such as:

- Heredity

- Nutrition
- Specific actions of the stressor

It is mainly due to such differences in conditioning that, through the same mechanisms, exposure to stress can produce varying hematologic changes under different circumstances.

### 6.2.2 Adaptability of Dairy Breeds to effects of Global Warming

Most dairy cattle in different parts of the world are normally subjected to climatic stress, which leads to a reduction of milk production. The level of production for most dairy cattle in tropical countries is considerably lower than the average production of similar cows in temperate countries. While much of the blame for this may be due to poorer management, a considerable portion might be due to the effects of the environment. This could be direct stress on the animal and by indirect environmental effects on fodder crops or other causes. It may be possible to recommend changes in management practices which would reduce stress and thereby possibly increase production. It may also be possible to develop indices of adaptation and selection for breeding for milk production.

Most *Bos taurus* dairy breeds are able to acclimatize where the mean annual temperature does not exceed 65°-70°F (18.3°-21.1°C). Most governments, especially in tropical African countries have tried to improve milk production by upgrading the local indigenous Zebu breeds (*Bos indicus*). It has been shown that with the increasing demand for milk, most local indigenous Zebu breeds would never be able to supply the needs in tropical African countries. Therefore, experiment with the introduction of more highly productive *Bos taurus* dairy animals is required.

The physiology of an animal may be affected by the environment both directly and indirectly. The most important direct effects are air temperature, solar radiation, temperature, humidity (relative and absolute) and air-flow.

Heat tolerance has been studied intensively in the last twenty years due to the importation of *Bos taurus* breeds of cattle from temperate zones into the tropics in an attempt to improve human nutrition. Work has been done both in the field and in climate laboratories, to determine the physiological effects of a tropical environment. Ariel & Robert, (2011) reviewed the whole field of the physiological effects caused by a hot environment on cattle. In tropical countries, the most acute problem for the animal is to keep its body temperature from rising beyond physiological limits.

There are several strategies that can be employed in order to alleviate the impacts of global warming and climate change on dairy farming. Some of these strategies include:

- Adopting a diverse and flexible feeding program for dairy cattle. This is advantageous especially in the case of erratic rainfall patterns in a place. For instance, dairy farmers can shift towards the use of annual crops and pastures where there is a problem of persistence pasture.
- Farmers can use sprinklers and shades to reduce the impacts of heat stress on dairy cattle.
- Ensuring there is plentiful water supply on the farm in order to meet all the demands for the livestock. Water drinking requirements can significantly be reduced by the election of shade on a dairy farm.

## 6.3 THE EFFECTS OF GLOBAL WARMING AND CLIMATE CHANGE ON BEEF FARMING

Beef farming practices are mainly done on rangelands. These rangelands are characterized by rainfall whose amount, distribution and reliability combine to create a very fragile environment. In most places, especially the tropics, the total annual rainfall rarely exceeds 800 millimeters.

Currently, there are two major beef production systems been practiced in the world and discussed briefly below.

### 6.3.1 Traditional Pastoralism

Such pastoralists keep beef cattle primarily for subsistence. Their husbandry practices are, by and large, those dictated by a subsistence economy. For instance, little or no capital has been invested in acquiring better animals or in improving management practices so that herds have to depend on the whims of the natural environment. In the past, the system presented no real problem as the indigenous cattle owners were able to roam over large expanses of the country and they, apparently, maintained equilibrium with their habitat. This is no longer the case as there has been a strong tendency towards settled livestock keeping.

**Figure 6.2:** Traditional pastoralism beef farming.

**Source:** https://en.wikipedia.org/wiki/Pastoralism

As the indigenous communities still depend on milk for subsistence, there is a tendency to maintain a high proportion of milking cows at all times, often in direct conflict with the largely seasonal availability of pasture. This practice causes huge losses of animals during prolonged droughts caused by global warming. For instance, it has been estimated that African countries alone (eco-zone IV and V) loss as much as 76% of its beef cattle population every year due to drought conditions (Ariel et al, 2012). All things considered, the off-take rate from pastoral areas is very low, estimated at 12%. It should, therefore, be considered a policy of most governments to draw the communities in these areas into the mainstream of national economic development through the provision of packages of appropriate technology based on scientifically generated information. This objective should seek to promote the development of the less favored areas.

## 6.3.2 Commercial Ranches

Commercial ranching is a common system used by large-scale beef farmers in the world. It is characterized by considerable capital invested in the acquisition of productive stock, fencing, watering points, dips, crushes, and mineral supplementation among others. These ranches contribute significantly to the supply of slaughter cattle in the country, especially to meat commission factories. In developing countries like Africa, the owners of commercial ranch properties were originally Europeans who were sufficiently knowledgeable in animal and ranch management practices. However, a number of commercial ranches have changed hands and are now owned by local citizens either individually or in partnership.

**Figure 6.3:** Commercial ranch.

**Source:** https://informationcradle.com/kenya/ranching-in-kenya/

The new owners have proceeded to split the ranches into smaller and smaller units and, whereas in the earlier situation, the large-scale rancher depended on sales of cattle for revenue, the small or medium-scale farmer undertakes production of food crops, milk and beef for both subsistence and sale. Whether sub-divided or not, the new ranch owners have tended to be unfamiliar with the practices of good livestock and ranch management with deleterious effects not only on animal productivity but, even worse, on the environment (Ariel et al, 2012).

### 6.3.3 Effect of Global Warming on Beef Farming

The livestock industry, especially beef farming, is viewed as taking very much of an economic second place to cash crop agriculture in the primary production sector. However, different climatic conditions have a serious direct influence on cattle distribution in most parts of the world. Of climatic factors, rainfall reliability is however considered by some to be an influential factor in delimiting beef farming.

Mean annual temperature and mean annual rainfall have a high relation to the presence or absence of cattle in any part of the world. Some researchers have laid stress on reliability and similar concepts concerning the periodicity of rainfall, in separating the arable and pastoral economies. Priority is often given in these rainfall reliability divisions to the climatic needs of arable agriculture which are considered to be more exacting than those of traditional pastoralism, which can thrive in a wider range of physical conditions.

Most parts of the world are classified as very arid (receiving less than 250mm rainfall four years in five), arid (receiving between 250 and 500mm rainfall four years in five), or semiarid (receiving between 500 and 750 rainfall four years in five). These zones correspond very closely to the ecological zones VI, V and IV. The climatic conditions of these zones can be attributed to global warming issues experienced worldwide.

Most of these zones are characterized with solar radiation, temperature and wind conditions unsuitable for beef cattle. Furthermore, rainfall is a major climatic factor affecting rangeland productivity. Examination of the patterns of rainfall in the semi-arid region would, therefore, be useful for providing a basis for consideration of likely variation in rangeland productivity, whether primary (vegetation) or secondary (animal).

One of the global warming characteristics of rangelands is the strong seasonality of rainfall with consequent prolonged dry periods. It has been noted that during droughts, the growth of pasture for beef cattle virtually ceases and grazing animals have to contend with what is essentially standing hay. Crude fiber content normally increases after a rainy season which in turn reduces herbage digestibility and, hence, intake of energy. There are also deficiencies in phosphorus and sodium. Deficiencies in chlorine, cobalt, and copper have also been reported in other range areas. Apart from mineral requirements, animals require energy and protein for body maintenance, activity, reproduction, production and growth.

A substantial part of an animal's diet is used for supporting body processes which must proceed regardless of tissue or milk production. If this need is not met, tissue breakdown occurs, resulting in weight loss, reduced conception rates, and reduced offspring weight at birth and weaning (Manjit & Surinder, 2013). The main consequence is that beef cattle in rangelands suffer from a periodic deficiency of both energy and protein which is often compounded with shortages of several mineral elements.

Seasonal deficiencies in quantity (undernutrition) and quality (malnutrition) of pasture bring about body weight changes which have been associated with low reproductive efficiency (Manjit & Surinder, 2013).

**Figure 6.4:** Emaciated beef cows due to the effects of global warming.

**Source:** https://medium.com/age-of-awareness/pastoralism-in-the-age-of-climate-change-adeea60553ca

Research has also shown that peak conception period coincides with the onset of rains. This implies two things:

- Firstly, there is no particular advantage to be gained by all-year-round mating other than the possibility of getting breeding females into calf with a minimum of managerial skills
- Secondly, mating would have to be organized when nutritional conditions are most favorable, (that is during the period of vigorous pasture growth)

The role of nutrition during the breeding season is influenced by both rainfall and temperature. For instance, high conception rates have been observed following the flush of good pasture during summer rains which confirms that conception is favored during hot months. Therefore, the low fertility levels observed during cold periods can be attributed to a supply of nutrients from pasture which is sufficient to bring about ovulation but inadequate to trigger off estrus activity. However, it should be realized that the seasonal effects of pasture would vary in various environments and the resulting effects on fertility would also be expected to vary.

### 6.3.4 Environmental Constraints in Beef Production

The beef industry is characterized by a low level of productivity which has been attributed to a combination of factors ranging from inferior stock genotypes, undernutrition, the prevalence of diseases and parasites, to sheer lack of managerial competence. However, the biggest challenge facing the beef industry is climatic conditions attributed to global warming.

**Figure 6.5:** Beef cows are more exposed to the total impact of global warming.

**Source:** https://newrepublic.com/article/121558/what-climate-change-doing-texas-cattle-ranch

Climatic conditions have been aggravated by periodic burning of bush, ostensibly to maintain grassland for grazing purposes, which has stripped vast areas of their vegetative cover. The problem has also been compounded by the part played by game animals in competing with cattle for the available grazing. Taken together, these factors have crystallized to make rangelands particularly vulnerable to environmental degradation which may often be irreversible. While dairy cattle are protected from the effects of global warming by the provision of shade, improved leys, supplementary feeding and veterinary care, beef cows are more exposed to the total impact of global warming such as excess heat and a shortage of water. Under these conditions, the important traits required are the ability to survive and reproduce. But mere survival and ability to reproduce are not enough. The beef industry is expected to play a bigger role in the overall economic and social development of a country instead of being relegated to a function secondary to crop production (Veerasamy et al, 2012).

The efficiency of production per se is not likely to make a major impact on the overall beef production required to meet the demand. A substantial proportion of the additional meat requirements would be expected to come from increased numbers of better quality cattle. Needless to say, direct importation of improved beef cattle to satisfy current and future demand is completely out of question. For the moment, the beef industry must primarily depend upon the maximum utilization of the adapted indigenous breeds as a basis for increased beef production.

To multiply the beef population quickly, it is critical to ensure a high level of reproductive performance by the breeding herd. A low reproductive efficiency not only reduces productivity but also limits attempts to improve the herd genetically. For instance, when calving rate is low, nearly all replacement heifers must be used to maintain the herd number and this curtails selection intensity thereby lowering the rate of genetic improvement. Besides, a decline in fertility would result in a compensatory increase in the number of unproductive and aged cows that are retained with adverse consequences of overstocking and soil erosion. Although several factors may influence fertility, there is overwhelming evidence that inadequate nutrition is the most important single factor improving the reproductive efficiency of beef cattle (Veerasamy et al, 2012). This is particularly relevant in tropical rangelands where pronounced seasonal changes contribute to a very precarious feed supply. It is now widely recognized that seasonal changes in feed supply are often reflected in the live weight of grazing cattle with adverse consequences on reproductive performance.

Therefore, body weight and weight changes could be used to monitor the fertility status of a grazing beef cow with a view to formulating nutritional and management strategies such as supplementation, planned mating, and strategic weaning in order to improve fertility. Furthermore, since nutrition, lactation, genotype, and management all interplay to bring about changes in live weight, a study of the variation in body weight in relation to these factors would provide a practical method of assessing their relative magnitude in the overall fertility complex.

Finally, evaluation of genotypes is necessitated by the diversity of exotic beef breeds already in the country and the need to assess the reproductive performance of crossbreds vis-a-vis the adapted indigenous breed types. Exotic beef bulls were introduced into Kenya for crossing with indigenous cows primarily to improve growth rate, milk yield and carcass grade (FAO, 2013).

Because of global warming conditions, it would be better to identify genotypes which can maintain a high level of fertility during prolonged droughts than to breed for improved growth rate and milk yield. Some of the problems and constraints mentioned above lend themselves to easier solutions than others but, all the same, constitute a big challenge to animal production scientists in their effort to increase beef production from rangelands. In this context, there is a need to set up minimum standards of management as a pre-requisite before formulating improvement strategies for rangelands. Such management practices are necessary, especially where there are improved stock genotypes and where husbandry practices are likely to respond to economic forces.

## 6.3.5 Nutritional Effects on Cattle Diets

Global warming and climate change have a marked influence on the quality, accessibility and availability of forage for the grazing animal. Forage selected by the grazing animal varies from that available. Pradeep et al, (2015) reported significant differences in forage species found in rumen samples as compared to the available forage species on the range, with the botanical composition varying between periods and to a lesser degree among sampling dates. Crude protein levels in the diets of grazing animals are reported to be significantly higher than those in available forage. These differences are caused by selectivity and preference for certain plant parts as well as certain plant species.

This selection process influences the nutritional quality of the diets. Decomposition apparently removes dead plant material, resulting in greater accessibility of green plant phytomass to the grazing animal as well as changing the species composition of the herbage. It has been noted that in vitro digestible organic matter from a spring paddock is greater than the central paddock due to a higher percentage of grass in the diets from the decomposition.

Season and rainfall are major influences on the growth and maturing or vegetation, and their major influences are reflected in the quality of animal diets. Dietary crude protein and *in vitro* organic matter digestibility (IVOMD) are reported to decline with the advance of season and maturity. Browse (leaves and twigs of trees and shrubs) and forbs generally contain higher levels of crude protein (and phosphorus) during the growing season than do grass and when available during the dry season. Browse crude protein content is more consistent across seasons. Thus, although browse generally has a lower digestibility than grasses, it provides supplements of protein and energy when the grasses are mature and of low nutritive value. Cattle have been observed browsing during herbage dormancy. The effects of browse are reflected in the crude protein, IVOMD and digestible energy (DE) of the diets (Pradeep et al, 2015).

## 6.4 GOATS FARMING

Arid and semi-arid areas around the world have been caused by the impacts of global warming. These areas form a great ecosystem for goats rearing for many pastoralists and semi-pastoralists. Approximately 45% of the total 1.2 billion people in Tropical African countries live in arid and semi-arid lands. These people are sometimes categorized as pastoralists and semi-pastoralists although the categorization is not exactly in correspondence to the aridity levels. They keep cattle and goats. Together, beef and small ruminant meat contribute 80% of the total meat supply of the Tropical African region, the other 20% coming from pork and poultry (FAO, 2013).

Both and goats play very important roles in the domestic supply of meat products. Among the roles include picking up the slack in domestic meat supply created by exports of live cattle and beef and being the choice slaughter stock for religious festivals. In addition, small ruminants are the choice of commercial slaughter stock in remote rural centers and homes.

In tropical African countries alone, the arid and semi-arid lands also referred to as rangelands, comprise 80% of the total land area of the region. These rangelands have been the major sources of beef and small ruminant (goats) meat for both domestic consumption and export. Exports of meat and meat products in tropics have been on the decline. Serious shortages in domestic supply of meat have been predicted over the last ten years. The reasons that these shortages have not been felt yet are probably due to lack of pasture and water caused by the impact of global warming (FAO, 2013).

**Figure 6.6:** Goat farming is also experiencing difficulties due to the impacts of global warming.

**Source:** https://phys.org/news/2017-06-chad-country-vulnerable-climate.html

The major producer of meat in Africa: South Africa had faced droughts in 2015 and 2016; that caused a lowering of livestock due to a decrease in a number of animals. Following the 2017 rains, the rebuilding of the herds that lowered the output of meat by 11.9% (FAO, World meat market overview 2017; 2018).The annual rainfall varies considerably across the world. This affects goat rearing practices in one way or the other. When sufficient rains are received, however, the rangelands recover very rapidly leading to a rapid buildup of herds and flocks.

### 6.4.1 Nutritional Effects on Goat Diets

The grazing animal is subjected to a continually changing pattern of food supply both in quantity and quality. At times, pasture growth is inadequate to meet even the animal requirement for maintenance, or there is deliberate rationing of the feed supply to conserve pastures for more critical periods of the production cycle. Forage is temporarily plentiful during and following the rains. The profitability then depends largely on the animals' ability to maximize intake. This makes the best use of the herbage before quality deteriorates. Systems based on breeding operations are generally more vulnerable to the changing climate due to the effects of global warming.

Information relating to the nutritive value of goat diets under range conditions is limited. Goats differ from cattle and sheep in their diet selection and gastrointestinal physiology. In the goat, the rumen is simpler but still contains food delaying mechanisms. However, goats have low enough metabolic rates and high enough gut capacity (compared to other browsers) that they can afford to selectively retain the more slowly digesting parts of plants (cell wall), but a small mouth and low absolute intake requirement permit selection for the higher quality parts of plants. Selective ruminant species may not have high pressure for large rumen fill as grazers, hence, they can select for rapidly digestible plant parts. Goats have faster rates of passage than cattle or sheep, which allows higher intakes, but less ability to digest fiber. The differences offer an explanation as to why the goats' elevated requirements can be satisfied under environmental conditions unsatisfactory for other animals, in that by use of their flexible upper lip, they can discretely select more nutritive diet constituents. For instance, the East African dwarf goats are not very selective in regards to plant species but are very selective to stage of growth and height above the ground of the plant material (Pradeep et al, 2015).

Goats are browsers and grazers. However, the literature on goats abounds with differing ideas concerning their feeding habits. It has been reported that browse provides the major portions of goats' diets. Most goats have a unique preference for shrubs and tree leaves. Diets of goats grazing a semi-arid consist largely of browse. The leaves are a consistent component along with a large portion of herbaceous material. However, goats feed on all major categories of forage depending on the availability and season. Thus, goats appear to have a superior adaptation to the dry tropics because of their ability to graze selectively and to willingly consume a wide variety of the vegetation (Pradeep et al, 2015).

Goats have a greater tendency than cattle and sheep to change their diets with changing seasons and achieving greater nutritional stability throughout the year. Values reported by Veerasamy. S, John. G, Lance. B and Cadaba P (2015) indicated that *in vitro* digestible organic matter of goat diets ranged from 43% in the late winter to 57% in mid-August, where the later peak was associated with higher amounts of grass in the diets. Dietary in vitro digestible nitrogen has been reported to be highest in sheep (1.26%), lowest in cattle (1.17%) and intermediate for goats (1.20%).

It has been noted that browse has more consistent crude protein content than grasses, which are typically high in protein at the beginning of the growing period and low in protein when they mature. Likewise, browse has higher lignin content than grasses, hence lower organic matter digestibility. Crude protein in leaves and young shoots of trees/shrubs is generally higher than that of herbs and remains high throughout the dry season.

**Figure 6.7:** Despite severe climatic conditions, goats are able to survive because they can browse as well as graze.

**Source:** https://www.dawn.com/news/1208617

Pradeep et al (2015) worked with Angora and Spanish goats and found the highest dietary crude protein values for Angora and Spanish goats grazing a live oak-shin oak savannah occurred when plant growth was most rapid (April) while the lowest values occurred in late winter (February). They reported the crude protein values to be proportional to the grass content and inversely related to the browsed content in the diets. He attributed the high crude protein levels in spring to the increased consumption of forbs and immature browse leaves. The low dietary crude protein values occurred when the proportion of dry grass in the diets increased.

## 6.5 POULTRY FARMING

Both of the environmental change and production of livestock have constantly negative effects one over the other. The environmental change could influence the production of livestock because of thc effects of expanding air temperature, feed-grain accessibility and favoring different illness. In Sub-Saharan Africa, the desire for environmental change will negatively affect the food security of various livestock. According to FAO (2013), production of livestock is one of the real reasons for the world's most pressing ecological issues. Even though much proof has been given on the negative effects of agricultural animal farming on global warming and other climatic conditions, this part has been overlooked by many researchers. Around 18% of negative effects of global warming mostly originate from the production of ruminant animals of which that ingests roughage and deliver methane gas. Lessening negative effects of global warming from the careful production of livestock as a rule and ruminants specifically are both urgent and basic (IPCC, 2014).

The poultry farming is portrayed by its industrialization, quicker development in utilization and trade than any other major agricultural livestock rearing practice on the planet. In any case, poultry production in most tropical nations depends on scavenging production.

Various factors enter into the production costs of poultry products. One of the major factors influencing the cost of producing meat and eggs is a shortage of poultry feeds. It has been estimated that the production of poultry feed is affected by the effect of global warming and climate change. Some of these effects include:

- Erratic rainfall
- Relative humidity
- Atmospheric temperature
- Pest and diseases

This has made the cost of feeds to increase in different parts of the world. The high feed cost component suggests the possibility of focusing efforts on reducing costs of production by reducing the cost of feed.

Meat birds of the hybrid type or broilers raised intensively and kept under optimal conditions reach slaughter weight within the prescribed time of 8 to 10 weeks. Intensive management of broiler-type birds begins with artificial brooding of the day-old chicks through the starter phase up to 4 or 5 weeks of age. During the later finishing phase and under tropical conditions, broilers do not necessarily require supplemental heat because physiological mechanisms are so fully developed such that the rate of heat loss to the surrounding is adequately compensated from the heat produced in the body. There are a variety of climatic factors that can influence the production of broiler chickens. These factors include:

- Atmospheric temperature
- Relative humidity
- Air movement
- Radiation

In view of this it is therefore important in broiler production that proper consideration be made to the control of these factors so that an optimal poultry environment be created at which broiler birds can benefit from better growth, increased feed efficiency and resistance to diseases, so that the ultimate product which is broiler meat can be efficiently produced. Although the floor area is considered a physical factor, it is very important as it influences many other factors, especially when there is high atmospheric temperature.

When there is high atmospheric temperature and chicks are overcrowded the loss of heat by the chick is greatly reduced. This results in a heat load on chicks which they cannot tolerate and therefore they appear to lose appetite for food and lose vigor which in turn leads to reduced resistance to diseases. Nutritional deficiencies due to a loss of appetite for food which affects feed intake and hence nutrient intake may also occur (Pradeep et al, 2015).

At high temperatures, chicks have the desire to dissipate heat to the surroundings to cool their bodies. In this case, they require less feed for maintenance and therefore consumed less feed. In other words, at lower temperatures, there is increased thermogenesis which requires more feed as a result of increased metabolic rates. James (1968) reported that at high temperatures feed intake decreases because of stress which results in a loss of appetite for feed plus the fact that at these high temperatures energy requirements for maintenance are less hence less feed is consumed.

Most of the studies on floor space requirements have been done in the tropical regions where adverse effects of global warming are significantly experienced, therefore there is justification for studies to be done in tropical regions to establish the optimal stocking rates at brooding and finishing phases under tropical conditions. In view of this lack of information, an examination has been made of the effects of different stocking densities during the brooding and finishing phases of broiler production in order to establish some guidelines for optimal floor space requirements for tropical conditions. Pradeep et al (2015) reported that if the production of broiler meat needs to be expanded quickly, it would be feasible in some instances to grow more birds per unit of space, despite the adverse effects from overcrowding.

Poultry farming has also been reported to cause global warming in one way or another. A lot of carbon dioxide is produced from the poultry farming sector. This gas originates from different fossil fuels utilized in this sector. For instance, a lot of propane gas is used in stationary combustion units such as incinerators and furnaces. Furthermore, a lot of diesel used in running various equipment such as tractors, trucks and generators also produce a lot of carbon dioxide gas. Another very common gas that is directly linked to global warming is nitrous oxide. This gas is emitted by poultry manure storage and handling process.

## 6.6 CAN GENETICALLY MODIFIED ANIMALS HELP IN ALLEVIATING THE EFFECTS OF GLOBAL WARMING?

Genetic modification involves altering the genetic structure of the animal by removing, adding, or changing the gene structure of the DNA sequence using artificial ways. It involves inducing certain traits of the organism and removing the undesired ones. The alteration may include induction of some characteristics that enhance faster growth and resistance to diseases. The changes induced to the animal using this technology can, therefore, be passed on to the next generations. The technology is possible; however, it requires a scientist to transfer genes that are important from one organism to the other animal of the same species. The technology has been used to modify different animals to have the desired characteristics like color, size and behavior. These animals include chicken, goat, cattle, dog, sheep, rat, mice, and pigs. The gene cloned from the mother animal should have the same gene strands to the animal that is to be modified (Ellen, 2011).

There are several technologies of incorporating DNA to the host gene and include:

- The use of biotechnology that has facilitated overcoming the physiological barriers enabling genetic material exchange in the living organisms
- Electroporation technique that facilitates gene incorporation into the genome. By use of electroporation DNA enters the other genome that it is inserted to through pores by electric impulse. The pores can be created by using microscopic crystals.
- Incorporating gene (microinjection) where DNA is directly introduced to the genome. To ascertain that incorporation has

successfully occurred, the incorporated gene before transfer to the site of incubation, is attached to the marker gene.

Many terms are used to describe the technology of genetically modified animals like biologically made and naturally derived. During the early times of technology development, genetic materials were transferred from one animal to another. However, with advancement in technology, current modification involves the creation of a new gene by either deletion or manipulation of the already existing genes. Animals that have undergone through genetic engineering have undergone several DNA mitochondrial changes such as mutation induction by use of either radiation or chemicals. Animals that have been cloned are also regarded as genetically modified since they involve various planning and intervention during their creation. Somatic cells are used to generate cloning of animals. Cloning involves replicating a certain cell to produce an organism with similar characteristics. The three types of cloning include:

- Therapeutic cloning
- DNA cloning
- Reproductive cloning

There are several ethical issues associated with genetically modified animals. These ethical issued deals with the animal welfare; type of procedures used where surgical and sacrificial of some animals has to be made to enhance genetic engineering. For example, a male can be subjected to vasectomy surgery under anaesthetic conditions to induce pseudopregnancy to the recipient embryo in women and the offspring should be of the same genotype. In females, ovulation may be forced by application of certain hormones.

The second ethical issue is a large number of animals required. Many embryos that undergo genetic engineering do not survive and those that survive, very few bear the genetic requirements of the alteration. This implies that many animals have to be sacrificed to produce many animals that have the genetic requirements which are contrary to society to minimize animal usage. This technology has recently increased the production of a variety of animals like mice. The technology has bccn rcdcfincd to facilitate the process of modifying organism; however, the procedures followed remain to be ineffective.

The third social implication is the anticipated welfare concerns. The genetic engineering of animals is inefficient and unpredictable due to an inability of controlling the DNA integration sites. According to Ellen (2011), there is also an unpredictable outcome of genetic engineering due to the interaction of the induced DNA to the host gene. This may interfere with the genome by either removing or adding DNA which results in alteration of the normal homeostasis. The behaviour of animals as a result of this genetic engineering, are unpredictable and may lead to several side effects for example, genetic engineering may suffer some abnormality such as increased weight, changes in histology and placenta (Ormandy et al, 2011).

There are also benefits of the genetic modified such as cow is increased milk production. The animals are more environmental friendly. However, they do not sustain themselves in a mix environment like dry seasons. The invention of technology is not accepted in several communities. Some communities that believe all animals have been created by God will not recognize genetically modified animals. Therefore, these organisms may have little or no use in such communities.

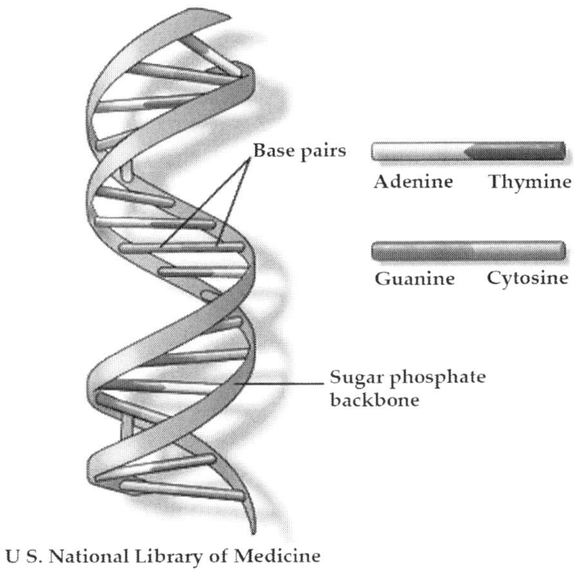


**Figure 6.8:** The DNA and RNA structures that show the interconnection of the four bases.

**Source:** https://ghr.nlm.nih.gov/primer/basics/dna

DNA stands for deoxyribonucleic acid. It is made up of several molecules known as a nucleotide. Every nucleotide contains a sugar group, a phosphate group, and a nitrogen base. The nitrogen bases are, thymine abbreviated as T, adenine abbreviated as A, Cytosine abbreviated as C, and guanine abbreviated as G. DNA instructions are expressed in by the order of the four nitrogen bases. A certain array of these bases forms a DNA sequence. Ribonucleic acid (RNA) and nucleic acids interpret the DNA information to proteins (Tutelyan, 2013).

Gene expression can be controlled by; regulating the transcription rate, RNA molecules need to be regulated in terms of stability . Rate of DNA transmission is supposed to be regulated.

Genetic engineered animals are prone to environments problems, according to some authors. They do not adapt to the environment easily in case they move to a new environment. However, new technology has enhanced vasectomy and pseudopregnancy but still; there is a problem with the procedures applied in conducting these processes of external fertilization. Due to the death of animals during the engineering process, few survive and the desired number of animals is not reached. The organisms that survive may come up with undesired characteristics than the intended ones. The genes transplanted from one animal to another may transmit diseases. These diseases can also be transmitted to human being indirectly (Tutelyan, 2013). For example, a disease transmitted to cows can affect human being through meat consumption. The introduction of genetic engineering has led to the absolution of indigenous animals that are very crucial. This can be seen in several ways such as cart pulling by bulls. The traditional bull shows to have an advantage over the genetically engineered bulls in terms of the energy to pull carts.

## Final note:

This chapter explored the effects of global warming on livestock such as cattle and goats. Certain benefits and challenges offered by genetic engineering were presented.

# 7 Relationship Between Impacts of Global Warming And Various Agricultural Pests and Diseases

CONTENTS


7.1 Introduction ........ 142

7.2 Livestock Ticks ........ 142

7.3 Brucellosis In Livestock ........ 147

7.4 Camel Trypanosomiasis ........ 147

7.5 Salmonellosis ........ 149

7.6 Cassava Mosaic Disease ........ 149

7.7 Cereal Storage Insect Pests ........ 151

7.8 Major Diseases And Pests of Coffee ........ 152

7.9 Cowpea Rust ........ 155

7.10 Major Diseases And Pests of Beans ........ 157

7.11 Plant Parasitic Nematodes ........ 159

7.12 Southern Leaf Blight ........ 160

## 7.1 INTRODUCTION

Both pests and diseases cause substantial loses in the agricultural sector. These losses are directed to both livestock and agricultural crops. These losses are caused by a wide range of arthropods like mites and insects. Furthermore, nematodes have been reported to cause huge losses too. Other pests include birds and rodents which are classified as vertebrates. Some of the major diseases that cause great losses in both livestock and agricultural crops are classified into fungal diseases, bacterial diseases, viral diseases, and microplasmas.

Losses caused by pests can be described in different ways. For instance, most pests cause or affect the crop in a direct way by reducing the growth which may end up reducing the total yields. Pests can also cause deterioration of yield quality. For instance, cotton bolls are deformed or discolored by pests. Pests of all kinds also lead to great financial losses, wastage of produce and also of land since more land is needed to grow the same amount of food if the yield per hectare is lowered by pests. To these losses must be added any losses taking place during storage and transport, for insects, rodents and fungi may destroy 10% or more of all stored produce in the absence of control measures.

In the developing countries, the pre- and post-harvest crop losses are estimated to be in the range of 30% or more of potential production. It should be noted that 30% of all world crops are lost to pests and diseases and it is worth about 75 billion US dollars. The loss of crops is estimated at 25% in Europe, 28% in Oceania, 29% in North and Central America, 30% in Russia, and communistic China, 33% in South America, 42% in Africa, and 43% in the rest of Asian countries (Kumar and Kalita,2017).

## 7.2 LIVESTOCK TICKS

Ticks are in the class *Arachnida*, and in the order *Acarina*. Economically, ticks merit special consideration for not only are they annoying pest, but in the temperate and tropical countries they surpass all other arthropods in the number and variety of diseases they transmit to man and his domestic stock. They act as an agent of transmissible infections of man and animals. Some of the most common diseases transmitted by ticks include Rocky Mountain spotted fever, East Coast Fever, South African tick typhus, Colorado tick fever, red- water, tularemia, heartwater, gall-sickness, looping ill of sheep, Nairobi sheep disease, encephalomyelitis, corridor disease, and many others (Christer & Niemela, 2015).

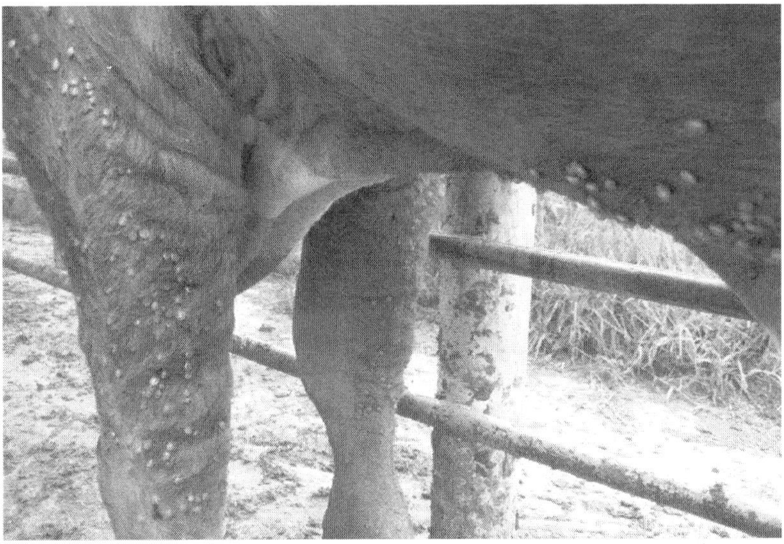

**Figure 7.1:** Ticks infestation on livestock.

**Source:** https://www.africanfarming.com/cattle-production-a-comprehensive-introduction-to-ticks-and-tick-borne-diseases/

Diseases caused by these organisms are the most widespread and are the major killers of livestock in many countries today. In tropical African countries, for example, about 50% of all cattle dying today do so from tick-borne diseases and 75% of these are due to East Coast Fever, a protozoan disease of cattle.

The life cycle of ticks consists of the following stages:

- Egg
- Larva: These are relatively smaller in size than either nymphs or adults. They have three pairs of legs and are sexually immature.
- Nymphal instars: They are relatively smaller than adults and bigger than larvae. They possess four pairs of legs, resemble morphologically the adult females but lack the "porose areas" which are found on the dorsal side of the basis capituli. They are also sexually immature.
- Adult: They are bigger than the other two stages, with four pairs of legs. The male and female have distinct morphological and physiological features. Sexual maturity in some ixodids is attained after a blood meal, while in others; sexual maturity is

attained immediately after molting even before taking a blood meal, e.g. some of the members of % the genus Ixodes.

The distribution of each tick species is influenced mainly by such climatic factors as:

- Relative humidity
- Temperature
- Vegetation
- The relationship between the tick species with the host species

For instance, most tick species are found in regions that have adversely been affected by the impacts of global warming and climate change such as arid and semi-arid. In situations where there are marked seasonal differences, there may be long periods when certain species of ticks are not to be found on their hosts. In tropical climates, seasonal changes are far less marked although they do occur. In these circumstances, the adults, in particular, of the various species of ticks are likely to be found on their hosts throughout the year (Christer & Niemela, 2015).

Although adults of ticks are active all the year round, their numbers markedly increase during the rainy seasons. In countries with only one rainy season per year, activity remains high throughout the rainy season. This pattern of seasonal occurrence has been shown in such countries in Central Africa like Malawi, Zambia, and Rhodesia. The main period of activity in these countries is between November to March each year with low numbers during the other part of the year which is generally hot and dry. In South Africa, the adults are active mostly between summer and autumn. In Kenya and Uganda, the adult population shows two marked peaks of activity, with both coinciding with the two rainy seasons of March- May (long rains) and October-November (short rains).

Therefore, the distribution of rainfall throughout the year must be taken into account. Generally, the most important factors affecting tick numbers and the variation in populations and species are: (i) rainfall distribution, (ii) vegetation cover (tied with stocking rate), (iii) total annual rainfall, and (iv) types of husbandry, vegetation and density of human populations.

**Table 7.1:** Seasonal incidences of some species of ticks

| Tick species | Seasonal incidences |
|---|---|
| *Ixodes ricinus* | This species has two phases of activity, especially in<br><br>Britain during the warm part of the year.<br>The first phase in spring (March-May) being the major peak of activity and the lesser peak of activity in autumn (September - October)<br>During the hot part of summer, there is a marked decline in activity.<br>In northern Russia, the tick is active in summer while in North Africa the tick is active in winter.<br>The most favorable temperature for tick activity is between 14-24°C<br><br>If the macro-climatic temperature is too high and humidity too low for too long during day time and suitable conditions are occurring deep down in the vegetation mat the ticks might then tend to remain immobile<br>On the other hand, where the micro-climatic gradient is steep, ticks may find suitable conditions at a point not too far from the tips for them to remain active and ready to attack.<br>The availability of the tick population at any given season will depend primarily upon two factors. First, on the timing of the cycle of development. Secondly, on the behavior and possible survival of unfed ticks from season to season.<br>In Russia, the seasonal activity regulation is ensured by egg, larval and nymphal diapause. In long day conditions (18-24 hours of light), engorged larvae molt into nymphs without delay, but in short day conditions (9-14 hours) diapause appears in all larvae and metamorphosis extends for about 200 days.<br>For unfed tick *Ixodes ricinus*, light is of very little importance and is never attracted to it and may be repelled by directed illumination, especially for newly molted ticks.<br>Unless permanently sheltered by a humid environment the survival of the terrestrial arthropod will depend to a considerable extent on its ability to limit evaporation. This is true especially as the fasting tick must often spend a long time awaiting a suitable host.<br>A tick is capable of taking up water from a humid environment or humid air. Similarly, it is capable of losing its water through evaporation to dry environment. As a result of this knowledge, several people have tried to relate tick activity to the water regulation.<br>The behavior of active unfed ticks is determined by many external factors of the environment and their physiological states. Tick climbing on the vegetation is ensured by their negative geotaxis but in unfavorable conditions of temperature and humidity, which reduce their normal water content, a positive geotaxis develops and the ticks return to the ground litter.<br>Direct reactions to the humidity gradient in air layers above the soil are also important. For instance, changes in the moisture equilibrium of the air, within the narrow limits permissible for Ixodes ricinus would not be expected to produce immediate responses, but rather to take effect through their prolonged application altering the water content of the body. |

| | |
|---|---|
| Hyalomma asiaticum | In tropical regions, the main adult seasonal incidence is observed in March-May, although all stages are found on hosts and in their principle biotopes throughout the year.<br>During the low activity of June-August, most adults are greatly emaciated.<br>Summer heat and increased saturation deficiency near the soil surface undoubtedly lead to the death of most ticks that have been unable to find hosts.<br>Spring engorged nymphs molt during summer into adults which remain in burrows until the following spring.<br>A few may become active and attack domestic animals in the fall of the same year.<br>Late summer engorged nymphs probably molt into adults during the same season. Nevertheless, if they engorge in the fall, low temperatures may delay molting until the next spring.<br>Therefore, the seasonal incidence of *Hyalomma asiaticumis* under the direct influence of climate and the presence of active unfed ticks depend on vertebrate host accessibility and on adult and immature survival for many months.<br>Before sunrise ticks are inactive, remaining motionless on the soil surface or hiding cover.<br>As air and soil surface temperatures increase at sunrise the ticks begin to move.<br>Maximum activity is observed when soil temperature rises to between 30-40°C and relative humidity is between 20-50%.<br>As the temperature begins to increase, ticks again begin hiding in various shelters.<br>By 1800 hours when the soil temperature begins to drops, the ticks again begin to be active.<br>At nightfall they are again motionless even when the air and soil surface temperatures are again high. |
| Dermacentor varlabilis | This has one major peak of activity during summer (June), with the activity spreading from March-August.<br>Researchers have shown a significant relationship between daily adult questing and daily solar radiation received.<br>Long term weather trends are also found to influence the activity of the tick populations.<br>More recently, it has been shown that questing behavior of the tick is regulated by its water balance.<br>A loss of 4-5% of the saturated equilibrium weight reverses questing behavior. |
| Ambyomma americanum | Both temperature and relative humidity gradient have a marked effect on the seasonal behavior of adult ticks.<br>The high temperature in combination with low humidity during day-time hours cause a gradual decline in total numbers of active ticks from early to late summer.<br>This is in response to an upper-temperature limit.<br>Increasing temperatures are very important in the initiation of early seasonal activity.<br>The high temperatures are responsible for higher metabolic activity and excessive water loss and that in many cases the water loss is the period when the relative humidity in habitats is above the critical equilibrium of the tick. |

| | |
|---|---|
| Amblyomma variegatum | This is essentially an African species, occurring in East, Central, Western, and Southern Africa and also in southern Sudan.<br>In Central Africa, with only one rainy season the species shows one peak of activity, while in East Africa where there are two rainy seasons they show two distinct peaks of activity.<br>The peaks always fall in the rainy seasons of April-May and November.<br>The tick species occurs at altitudes from sea level to at least 8,500 ft. (2,800m). |

**Source:** https://www.nature.com/news/the-growing-global-battle-against-blood-sucking-ticks-1.18227

## 7.3 BRUCELLOSIS IN LIVESTOCK

Brucellosis is a worldwide zoonosis which causes abortions and infertility in animals and debilitating disease in humans. The disease is of public health significance because it can directly or indirectly be transmitted from infected animals to man with consequent illness and prolonged incapacity. It is also important in animal production because it causes serious losses which arise from abortion, premature birth, infertility and decreased milk yield. High frequency of brucellosis is found in areas that have been adversely affected by the impacts of global warming and climate change such as arid and semi-arid. In these areas, extensive cattle husbandry is practiced and the cattle population is large (Pradeep et al, 2015).

Brucellae are Gram-negative coccobacilli without capsules, are non-motile, non-spore forming and do not show bipolar staining. They are strict aerobes, but some require 5-10$ added carbon dioxide ($CO_2$) for growth especially on initial isolation, the temperature range for growth is 20 to 40° C, and pH range is 6.6 to 7.4. High temperatures in arid and semi-arid areas favor their growth and development.

## 7.4 CAMEL TRYPANOSOMIASIS

Camel production is concentrated in an area that has been adversely affected by the effects of global warming and climate change. These areas include:

- The desert areas
- The arid areas
- The semi-arid region

These camel keeping areas are inhabited by nomadic pastoralists. Camel milk and meat is a source of protein, and in addition, camel milk, meat and hides can be sold for cash or exchanged for other goods. Camels are the major transport animals in these areas and may well remain as the only means of transport in this region for a long time. As a result of the significant growth in world camel production and the current increase in desertification because of global warming and climate change, the economic importance of camel health should be recognized (FAO, 2013).

**Figure 7.2:** Camel are severely affected by trypanosomiasis due to effects of global warming in desert areas.

**Source:** http://moonspaghetti.blogspot.com/2011/10/lake-turkanaomo-valley-route-from-kenya_4700.html

Trypanosomiasis, which affects camels, has had less scientific attention. In view of reported outbreaks of camel trypanosomiasis worldwide, coupled with the fact that the mode of transmission of camel trypanosomiasis is not clearly understood, there is a clear need to intensify research on this disease. Knowledge of peak periods when camels are challenged most by trypanosomes can be useful to control campaigns and targeting of chemotherapy at the most appropriate time (Veerasamy et al, 2012).

Some of the major climatic factors attributed to the trypanosomiasis infection on camels include:

- Temperature
- Relative humidity
- Wind
- Dust

These factors influence the survival rate of trypanosomes in the fly (vector) mouthparts. In Africa, trypanosomiasis is prevalent in Morocco, Algeria, Tunisia, Chad, Libya, Mali, Egypt, Senegal, Sudan, Somali and northern Kenya. The infection has also been reported outside Africa in places such as Israel, Lebanon, Syria, Arabic Peninsula, Asia, Iran and Iraq. In Europe, the Transvolga region (except Turkey and Bulgaria) is an area chiefly affected by trypanosomiasis infection in camel.

## 7.5 SALMONELLOSIS

Salmonellosis is a public health hazard. It is directly linked to the production of poultry. It causes severe gastroenteritis and septicemia in animals. These conditions may lead to death. As a result of the septicemia, abortions, polyarthritis, endocarditis, nervous disorders and other conditions may also arise. Poultry salmonellosis is economically important to the poultry keeper. The disease may cause high losses by death during the first three weeks after hatching. It can also cause reduced fertility and hatchability. It also may cause reduced egg-production and increased mortality among adult flocks. The most damaging effect of Salmonella infection on young chicks is one of stunted growth to the extent that infected birds are of low quality when they reach market age.

The organisms also serve as secondary invaders in many avian diseases. Although the optimum temperature range for growth of Salmonella is 35 to 37 °C, Salmonella strain possesses an unusual heat-resistant property. To inactivate it, about 55° C is required. In the environment, salmonellae can remain viable for a long time. Survival of 87 days in tap water, 115 days in pond water, 120 days in pasture soil, 280 days in garden soil, over 30 months in dried bovine manure, 28 months in naturally infected avian feces and 47 days in manure slurry have been reported. It can therefore be concluded that effect of global warming such as adverse temperature and humidity may not stop its effectiveness on poultry (Veerasamy et al, 2012).

## 7.6 CASSAVA MOSAIC DISEASE

Cassava mosaic disease distribution is worldwide primarily due to the use of infected planting material, the widespread presence of the vector and the use of traditional local varieties which seem to be susceptible to mosaic. It was first reported in Africa in 1894. Total development period of the disease in the tropics ranges from 11 to 50 days depending on temperature and relative humidity. It produces 12 generations in the course of one year. It has a wide host range including both wild plant species and cultivated crops amongst which are cassava, cotton, cowpeas, peppers, sweet potatoes, tobacco and tomatoes. Furthermore, the seasonal activity of the whitefly depends on temperature and light. Whiteflies fly away from cassava in the morning (Ariel et al, 2012).

**Figure 7.3:** Whiteflies.

**Source:** https://phys.org/news/2017-09-africa-notorious-insects-bugs-agriculture.html

Rainfall, relative humidity, solar radiation and temperature on population development of cassava mosaic disease accounts for about 50% of the variation with 45% due to atmospheric temperature and relative humidity.

Significant increases in numbers of whitefly subsequent to rainy periods with major peaks during the long and short rains have been observed. Dry months are characterized by a gradual reduction in whitefly population density. These fluctuations may be explained by the effect of rainfall on the host plant resulting in the production of young succulent leaves which are necessary for reproduction of the whitefly. In addition to the indirect effect of rainfall on whitefly population, it has also been observed that relative humidity and temperature interaction is highly correlated with the development of cassava mosaic disease in the field (Ariel et al, 2012).

## 7.7 CEREAL STORAGE INSECT PESTS

Cereal storage insect pests belong exclusively to the order *Coleoptera*. These pests have been reported infesting stored cereal in tropical countries where there are favorable weather conditions such as high temperatures. A pest like *Callosobruchus maculatus* infests cereals while it is still in the field just before harvesting so that the seeds coming into the store contain immature of the pest, and the population of the pest builds up soon after the commencement of seed storage (Moshe & Wajnberg, 2017).

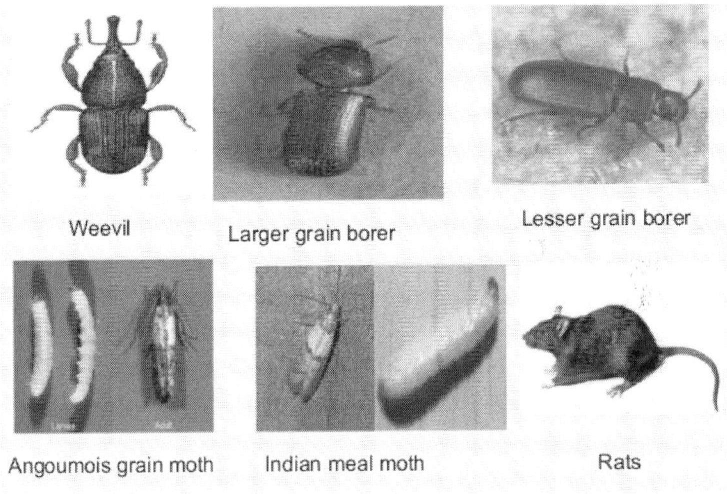


**Figure 7.4:** Major pests of stored cereals.

**Source:**https://www.slideshare.net/20107-07/seeds-visit-seminar-june-2015-insect-pests

On the temperature conditions of storage, studies have indicated that the rate of population buildup of storage pest especially *Callosobruchus maculatus* comparatively much lower at a lower temperature. At this low temperature, there is the least number of *Callosobruchus maculatus* yielded by stored cereals

Based on these observations, it can be concluded that the storage of cereal at lower temperatures can be used by farmers to reduce the pest population increase and subsequent damage to their crop. The construction of well-aerated granaries that keep cool during the hot season would be recommended for cereal storage (Moshe & Wajnberg, 2017).

## 7.8 MAJOR DISEASES AND PESTS OF COFFEE

The annual consumption of coffee stands at 400 billion cups globally; it is one of the legally traded items of high value with millions involved in production, processing and sale (Jha et al, 2014). Coffee planted in areas which are too warm is more susceptible to diseases. The most serious coffee disease is the coffee berry disease (CBD), which appeared for the first time in 1922. This disease attacks the green tissue of coffee berries at any stage of their development, often penetrating into the inner rim and destroying the beans. This disease spread quickly. Losses attributed to CBD increased rapidly each season (David & Howden, 2013).

Coffee berry disease is caused by fungus *Colletotrichum coffeanum*. This fungus attacks flowers, green berries, ripe berries, and sometimes leaves of young suckers. The classification of coffee berry disease is based on the growth of the berries and the type of parts attacked. Coffee berry disease affects immature coffee berries. The susceptibility of the young berries increases as they expand during the period of 4 - 16 weeks after flowering.

**Figure 7.5:** Coffee berry disease.

**Source:** https://www.flickr.com/photos/44393503@N07/4100779206

"Brown-blight" is a term used when the pathogen attacks ripe coffee berries. "Berry scab" is a term used when the development of the active lesions is halted by the onset of dry conditions and the berry may be the only part affected. When the disease attacks the leaves, it is referred to as "leaf anthracnose" and if on branches as anthracnose of the branches. The disease has now spread to most African countries but has not yet been reported outside of Africa. However, a coffee berry necrosis that was first observed in Guatemala in 1963 is present in Costa Rica and may be similar.

There are several components of global warming closely associated with germination and infestation of coffee berry disease. Temperature is one of these components. The optimum temperature for the germination of *Colletotrichum coffeanum* conidia in water is about 22° C, but the presence of nutrients results in a higher optimum about 27° C although the optimum for subsequent lesion formation after inoculation remain at 22° C. The maximum temperature for germination is about 30° C (up to 35° C with nutrients) and the minimum is 15° C. However, it has been shown that some infection could occur at temperatures below 15°C (David & Howden, 2013).

Another very significant component of global warming that is associated with the infection is relative humidity. Germination occurs only in the presence of moisture and high humidity. Meteorological data from within coffee trees shows that at temperatures allowing germination and infection, a saturated atmosphere in the absence of free water would be rare. To curb the threat posed by CBD, effective fungicides have been developed and this disease is now under control.

**Figure 7.6:** Coffee leaf rust.

**Source:** https://santaram09.wordpress.com/2010/01/08/coffee-leaf-rust-resistance/

The other major coffee disease is known as Leaf rust. It attacks coffee grown in low altitude warm areas. The fungus attacks coffee leaves in the wet season making them fall off the tree. This reduces the trees nutrient formation process and results in disappointingly low coffee yields. Leaf rust was particularly serious in the 1950 s. Consequently, Leaf-rust resistant coffee strains were developed and at the same time, effective fungicides were discovered. The disease has therefore been largely brought under control (David & Howden, 2013).

Coffee is sprayed with fungicides eight times in a crop year. Spraying of fungicides begins just before the rains start and is intensified during the rainy season; ending when the crop is ripening.

Coffee is also attacked by numerous types of pests. Most infestations are influenced by prevailing weather conditions. The most serious pest of coffee trees is mealybug. Initially, mealybug was controlled by a technique known as "cresto-banding". Later more effective banding greases were discovered and now mealybug is under control.

**Figure 7.7:** Coffee mealbug infestation.

**Source:** https://www.greenlife.co.ke/mealybugs/

Thrips is another pest which posed a serious threat to coffee in the 1940s. A few years later, however, this pest could be controlled easily by insecticides which were then available.

Antestia bug and Leaf miner have also been a menace to the coffee industry. Fortunately, these pests do not worry coffee farmers now as much as they did in the 1940s, because very effective insecticides have been discovered to control them. Due to sporadic pest attacks on coffee attributed to different climatic conditions, recommendations for insecticide use are not very precise, especially for the frequencies of insecticide application in a crop year. Farmers are simply advised to spray an insecticide whenever they detect pests on their coffee. The quantity they use and the number of applications within a crop year, however, depends on what they consider to be the economic level of insecticide use. To effectively control coffee diseases and pests, and hence maintain high coffee quality, the government should form a statutory board, charged with the responsibility of supervising coffee production activities, including licensing growers, coffee millers, nurseries and marketing (Christer & Niemela, 2015).

Another common pest of the coffee are fruitflies. There are different species of fruit flies of the Family *Trypetidae* commonly breed in mature ripening coffee berries in tropical countries. Three major ones include:

- Ceratitis capitata Wied
- Ceratitis rosa Karsch
- Ceratitis coffeae Bezzi

Their larvae live in the coffee berry feeding on the fleshy pulp or mucilage. Prematurely dropped coffee berries have been found bearing punctures caused by fruit flies. An extensive premature drop of coffee berries can considerably lower the yields.

The life cycle of fruit flies is well documented. The incubation period of the egg of *Ceratitis capitata* takes 2 to 4 days. Between 14 and 28° C, the larval period lasts 6 to 11 days with an average of 2 days between the instars. A total of 4 instars preceed the pupal stage. The larval stage takes 9 to 15 days under unspecified temperature and humidity conditions. It has also been observed that moisture does not prolong the pupal duration although it affects the survival of the pupae (Christer & Niemela, 2015).

## 7.9 COWPEA RUST

Cowpea (*Vigna unguiculata* L. Walp.) is cultivated over 12.5 million hectares in the dry savanna of the tropics with Nigeria as a world's largest producer of cowpea (Haruna and Usman, 2013). In addition, the green leaves and grains are sometimes sold for cash in the local market. Nutritionally, cowpea is very important. It is rich in protein and vitamins. The protein content of leaves is estimated at about 40% on dry weight basis while that of the dry grain is 25%. The crop contains vitamins such as thiamine, methionine, niacin, pyridoxine, pantothenic acid and folic acid.

Production of cowpeas is, however, affected by many factors among which plant diseases constitute an important factor. These diseases reduce the quality and quantity of the leaves and seeds considerably. More than ten diseases have been recorded on cowpeas. Cowpea rust is among the top five major diseases which are responsible for reducing crop grain yield appreciably. If the disease appears early, it completely defoliates the crop. Unfortunately, very little work has been done on the cowpea rust in Africa and therefore, no indication of the losses caused by the disease can be ascertained. Although the disease is of economic importance it has attracted little attention in Africa and many other parts of the world.

**Figure 7.8:** Cowpeas rust as a result of the adverse effects of global warming.

**Source:** https://plantvillage.psu.edu/topics/cowpea/infos

Some of the components of global warming and climate change attributed to cowpea rust infestation include relative humidity and adverse atmospheric temperature. Researchers have also reported an increase of the disease from 30% when rainfall was 40 mm to the maximum intensity when rainfall was 205 mm. This has led to the conclusion that the main environmental factors in the development of cowpea rust are relative humidity, temperature, and rainfall.

**Figure 7.9:** Adverse impact of global warming on cowpeas production.

**Source:** https://plantvillage.psu.edu/topics/cowpea/infos

Light has also been found to affect disease development. For instance, researchers have reported a reduction of infectivity with uredospores kept in the dark as compared to those kept in light and intermittent light respectively. Still, David and Howden (2013) observed retardation of rust development due to the absence of light during uredospore formation.

Use of resistant varieties has been found to be the most effective means of controlling cowpea rust. In most cases, this is the only method recommended for the control of this disease since chemical control is considered uneconomical for this particular crop.

## 7.10 MAJOR DISEASES AND PESTS OF BEANS

Diseases and pests are the major constraints to bean production in different parts of the world. Most of the diseases and pests that attack beans are influenced by different components of global warming and climate change. Some of these components include:

- Erratic rainfall patterns
- High temperatures
- Inadequate or unbalanced plant nutrients
- Poor soil conditions

Various diseases and pests limit the production of beans. Some of the major diseases of beans include:

- Fungal diseases - Ashy stem blight, powdery mildew, anthracnose, southern blight, rust, Fusarium root rot and angular leaf spot
- Bacterial diseases - Bacterial brown Spot, halo blight, common blight, and fuscous blight
- Viral diseases - Bean common mosaic virus

Insect pests attack beans both in the field and in stores. Several insects are of major economic importance because they cause considerable reduction and quality reduction in beans. The bean flies are by far the most important field pest of common beans. Bean aphids are serious pests of beans, especially in areas that have been affected by adverse impacts of global warming such as arid and semi-arid areas. They attack the foliage, tender stems and growing points of bean plants. They suck the plant sap from leaves and stems. Infested leaves are often cupped and later turn yellow, resulting in the death of the plant. Besides the direct damage to bean plants, the bean aphid serves as a vector of the bean common mosaic virus disease.

The African bollworm is a sporadically serious pest attacking the flowers and pods of bean plants. Young larvae feed on flowers and young pods but the main damage is caused by old larvae burrowing into the green pods and eating the developing seeds. Bean bruchid is the most destructive pests of stored beans. Infestation starts in the field and continues in the store if unchecked. Larvae feed inside the seed and make tunnels almost to the surface leaving the seed coat to form a window at the end of the tunnel through which the adult emerges. Beans which are attacked by bruchids are not only unfit for human consumption but are also unsuitable for planting.

There are a number of pests, considered to be of moderate or lesser importance, but which could become a threat to bean production (Lichtfouse, 2009).

These include:

- The blue beetle and green beetle: They attack the leaves of bean plants mainly in the early stages of growth, sometimes leading to the death of the bean plants. The larvae live in the soil and can damage the root system.
- Pollen beetles: They feed on the pollen and destroy the anthers.
- Blister beetles: They are common but minor pests of bean flowers. They feed on floral parts and destroy them.
- Bean thrips: They attack flower buds and flowers of bean plants. The feeding injury is characterized by distortion, malformation

and discoloration which lead to abortion of flowers.

- The Legume pod borers: They attack flowers, pods and seeds of bean plants. The early instars feed on flowers and later instars feed on pods. The presence of legume pod borers in the pod is characterized by frass around the entrance hole.
- Spring brown bugs: They suck sap from pods as they are formed, causing premature drying and shriveling of pods. The seeds remain small or they may not develop at all.
- The Green stink bugs: They attack pods, seeds and leaves. The feeding punctures cause necrosis resulting in deformation or pod shedding if the bean plants are attacked when very young.

Tropics and warm temperate regions have been identified as the most important field insect pests of beans in Africa. This is because of the weather and climatic conditions favoring multiplication of these pests in large numbers. Studies have also shown that the seriousness of these pests varies from season to season and they are generally most severe during the dry conditions when up to 100% losses have been reported (Lichtfouse, 2009).

The biggest damage is caused by the larvae feeding in the stem causing calloused growth around the injured base of the stem and leaf yellowing. The attacked seedlings may become severely stunted or die. Late infestation does not have serious effects on the plant. Various control measures are recommended for bean flies. These include:

- Pre-or post-planting application of various chemical pesticides
- Early planting
- Hilling up
- Host plants elimination

Bean cultivars resistant to bean flies, a potential means of controlling bean flies would be most welcome to the small scale farmers, especially since most of them cannot afford to buy insecticides and other control measures are more difficult to implement.

## 7.11 PLANT PARASITIC NEMATODES

Plant parasitic nematodes have been reported as among the most common and leading disease problems in agricultural crop production, especially in tobacco. The identification of 4100 species of plant-parasitic nematodes has been reported that together cause crop damage amounting to $80–$118 billion. Such nematodes target the roots to lower the yield and performance (Bernard et al, 2017).Climatic factors attributed to impacts of global warming that influence nematode infestation in crops include:

- Temperature
- Moisture
- pH

However, they are of secondary importance with regard to nematode damage on a susceptible crop since soil factors that will allow plant growth, will always allow nematode development.

Nematodes are usually controlled through an integrated approach which includes crop rotation, farm hygiene, resistant cultivars and soil treatment with nematicides. Small scale farmers usually eradicate nematodes from the top soil of the prepared seedbeds by burning a pile of firewood over it.

Due to the high susceptibility of tobacco to nematodes and to its similarly high value per unit of land, there has been a spectacular economic return from controlling nematodes on "problem fields". Under an integrated nematode control program on f tobacco farms in North Carolina, U.S.A., the following results have been reported:

- sanitary procedures such as plowing-up tobacco roots at the end of season, increased crop value above control plots by 68%,
- Rotation with resistant crops added 105%,
- Use of resistant tobacco varieties added 90%, and
- use of soil chemicals, particularly combined nematicides and insecticides, added 132% (Moshe &Wajnberg, 2017).

## 7.12 SOUTHERN LEAF BLIGHT

This disease has a worldwide distribution mainly in areas that have adversely been affected by the effects of global warming such as tropical and sub-tropical areas. In Africa the disease has been reported in Congo, Egypt, Ghana, Guinea, Ivory Coast, Malawi, Nigeria, Senegal, Sierra Leone, Sudan and Zambia while in Asia it has been reported in Cambodia, China, Hong Kong, India, Indonesia, Japan, Korea, Malaysia, Nepal, Pakistan, Philippines, South Vietnam and Thailand. Cyprus, Denmark, Germany, Italy, Portugal, Rumania, Spain and USSR (Caucasus) are the European countries in which the southern leaf blight of maize has been reported. In the Americas and West Indies the disease has been reported in Canada, USA,

Bahamas, Cuba, Jamaica, Nicaragua, Panama, Trinidad, Argentina, Bolivia, Colombia, Ecuador and Surinam, British Columbia, Solomon Island, Fiji, New Caledonia, New Hebrides, Samoa, Papua and New Guinea are the countries of Australasia and Oceania in which the disease has been reported. The southern leaf blight of maize was so named because in the USA it is found in the southern states of Iowa, Nebraska, Indiana, Illinois and Texas that form part of the Corn Belt.

Infection normally increases essentially in a straight line relationship with increasing dew and colonization temperatures and increasing dew periods. Fluctuating day/night temperatures retard lesion expansion as compared with a constant temperature near the mean of the day/ night temperatures. A lower relative humidity during colonization markedly reduced lesion size. Researchers in Florida in the US have observed that the southern leaf blight of maize become more severe on field maize when minimum temperatures are 20° C or above and relative humidity is 100%. This indicates that warm, moist weather favors southern leaf blight epidemics (David & Howden, 2013).

## Final note:

This chapter outlines various pests that affect agriculture and livestock; this becomes important in the context of global warming.

# Economic Importance of Irrigation in Agricultural Sector With Reference to Global Warming

# 8

CONTENTS

8.1 Introduction........164
8.2 Definition and The Need of Irrigation in Agricultural Sector........164
8.3 Major Sources of Irrigation Water........164
8.4 Benefits of Irrigation........165
8.5 Factors Influencing the Appraisal of Irrigation........167
8.6 Methods of Irrigation........170

## 8.1 INTRODUCTION

A large part of the world is classified as a semi-arid zone in that the rainfall is less than 500 mm per year, poorly distributed, and often punctuated by long dry spells. This type of rainfall regime naturally affects food production so that most of the foodstuffs in the word have to be supplied from rural areas. This could be done by expanding minor irrigation programs mainly through private investments in wells.

## 8.2 DEFINITION AND THE NEED OF IRRIGATION IN AGRICULTURAL SECTOR

Irrigation is the artificial application of water to crops, either to supplement or to replace rainfall, and thus to assist in creating optimum conditions for high yields (Chuet al, 2015). The basic objective of setting up an irrigation scheme is to maximize social welfare. Since welfare cannot be measured in real physical terms, proxies such as output, employment, equality and risk minimization are used in indicating welfare gain. The maximization of welfare is equivalent to the minimization of social costs.

Some regions have characteristics which reduce the social, technical and economic costs that are to be minimized when setting up an irrigation scheme. Irrigation is undertaken under diverse and complex situations of the environment and socioeconomic conditions of the people. Proper planning and coordination of decisions would require a model that is comprehensive and flexible enough to accommodate most of these diverse situations. The linear program model is flexible enough to accommodate these situations. Most of the irrigation schemes are required to keep records of the activities that go on in the scheme which reduce the cost of data collection over time.

## 8.3 MAJOR SOURCES OF IRRIGATION WATER

Water for irrigation is normally obtained from two general sources:

- Surface water: This is usually in the form of rivers, streams, and lakes and may be made available for irrigation by the simple diversion of the streams or rivers or by using pumping equipment.
- Groundwater: This occurs below the surface of the ground in a zone in which permeable rocks are saturated with water under hydrostatic pressure. Irrigation using groundwater involves pumping from a depth of a few meters to several hundred meters

except where artesian wells exist. This source of irrigation water in most areas requires a source of power, which may be manual or by animal or mechanical means

Despite the fact that in many parts of the world irrigation development has been extended to cover even the humid and sub-humid areas in the world, much of the irrigation development efforts have been concentrated on the so-called medium and low-potential zones. The main objective in concentrating efforts in these areas has been to mitigate the effect of global warming such as famine and drought and to bring security, stability, and prosperity to those areas hitherto producing only catch crops or no crops at all. Nonetheless, as modern cash inputs are successfully removing or reducing the effects of many limiting factors, even in areas of moderately adequate rainfall, there is an increasing need to consider supplementary irrigation to prevent moisture-availability from becoming the ceiling on yields (Chuet al, 2015).

## 8.4 BENEFITS OF IRRIGATION

Among the specific benefits that may accrue from irrigation are the following:

- Maximum growth and yield traceable not only to the supply of water itself but also to the part that moisture plays in making nutrient materials available
- The timing of maturity of crops - particularly desirable where earliness is profitable
- Maximum table and market quality (Quality of the harvested product is almost always inherently improved through irrigation)
- Utilization of land

Irrigation usually affords an opportunity for fuller employment of land by ensuring double or multiple-cropping.

One of the most important factors determining proper irrigation practices is the character of the soil being irrigated. In general, the texture, structure, and porosity of soil determine its water-retaining and transmitting capacity. In turn, these two capacities, together with the crop and depth of the root zone, largely govern the method of irrigation, the frequency of application, and the quantity of water that should be applied. The water retaining capacity of the soil, for the purpose of irrigation, is expressed as depth of water held in a given depth of soil. This is expressed in inches depth of water per foot depth of soil or millimeters of water per centimeter depth of soil. When the soil is thoroughly watered, some of the water, under the influence of gravity, drains into the lower levels and is replaced by air from the surface. When drainage virtually ceases usually after two to four days, the soil is said to be at field capacity. The roots of crops obtain water from the film of water held around the soil particles by surface tension. As this film becomes thinner and thinner, the roots find it increasingly difficult to take in water. When the roots can no longer take up water sufficiently rapidly to remain turgid, transpiration ceases and the plant wilts. The soil is said to be at the permanent wilting point. In a given soil, all ordinary plants wilt permanently at the same moisture content. Field capacity and permanent wilting points are the two important water-retaining capacities of soils as far as irrigation is concerned (Ouda, 2015).

For most crops, the reduction in yield becomes significant when the available moisture is below 50% for prolonged periods. In other words, the average growth of a plant in any period is a function of the average level of available soil moisture in that period. The effect of irrigation on yield is directly related to the amount of water held in the soil reservoir until the maximum yield is produced. Additional water will not increase the yield. Therefore, crop yields increase with irrigation water supply within a fixed range. Higher crop response to irrigation is realized with the optimum application of other inputs like fertilizers and manure. Organic manures, for example, modify the soil structure so that it can hold more water and air for the extra benefit of a crop.

Reliable information about when plants need specified amounts of water is required to permit scientific irrigation of crops. Most crops have two or three periods of maximum water requirement (Ouda, 2015).

- The first, during the seedling stage
- The second, during the pre-flowering and flowering stage
- The third during the seedling stage

Young seedlings are particularly susceptible to water tension, and growth is retarded if they are subjected to high water tension. For efficiency in water use, farmers need to be able to determine the proper time for irrigation. Three criteria can be used for this purpose:

- The soil becomes depleted of soil moisture which shouldt be replenished
- The soil also becomes unable to supply sufficient moisture for maximum growth and there is a cessation of plant growth followed by other indications of insufficient water

- A change in color of foliage and perhaps wilting

Farmers in developed areas determine the need for irrigation by sampling the soil with an auger and determining the approximate wetness by color or feel. Most soils change color between the field capacity and permanent wilting point; usually, the color is darker at field capacity and becomes lighter in color as the moisture content approaches the permanent wilting point. The color of foliage becomes dark green, often almost bluish or grayish, as the supply of moisture becomes insufficient. The quantity of water that should be applied at each irrigation and the frequency of irrigation depends on:

- The soil type
- The crop
- The weather
- The soil storage capacity
- The water already held in the soil
- The rate at which water is absorbed through the soil

Two important characteristics of the crop which affect the frequency and the rate of irrigation are the depth of the root system and the stage of growth in the life cycle of the crop. The rate of transpiration will vary from practically nothing at the young seedling stage during cold, cloudy weather to a maximum during hot windy weather at the time when crops are growing luxuriantly. Each crop has a certain, rather definite, rooting habit which it will tend to follow if the soil is uniform and deep enough and equally well-moistened. On the other hand, almost any crop will develop its major root zone in the most favorable environment with regard to both nutrients and soil moisture (Ouda, 2015).

## 8.5 FACTORS INFLUENCING THE APPRAISAL OF IRRIGATION

There are several factors involved in any appraisal of irrigation water. Among these are:

- The total quantity of dissolved salts
- The particular constituents and their ratios
- The characteristics of the soils and the crops to be irrigated
- The irrigation practices

- The climate particularly temperature and humidity

Many authorities agree that the suitability of water for irrigation purposes depends on the effects of its total quantity of dissolved solids on the plant and the soil. Salts may harm plant growth physically by reducing transpiration through modification of osmotic processes, or chemically by their toxic constituents (Bhupinder, Annette, & Chan, 2011). Salts affect soils by changing their structure, permeability, and aeration and this indirectly affects plant growth. Soils, particularly fine-textured ones, have the ability to absorb certain minerals from irrigation water. If irrigation water contains more sodium than it does calcium and magnesium, a tendency exists for this sodium to replace the calcium and magnesium already in the soil. The presence of excess sodium in the soil makes the soil less permeable, so the effect of irrigation water containing an excess of sodium is to tighten or seal-up the soil.

The productivity value of water in the industry is frequently 100 times or more as great as for agriculture. However, factors other than the direct financial return per unit of water used have in the past been considered to justify the existence of some expensive water-development projects. As capital is one of the scarcest resources in a developing economy, it is essential that economic and financial evaluation and appraisal of irrigation projects be done to determine the return to the scarcest resource. Capital and operating costs of providing irrigation and the returns that can be expected under different conditions vary widely.

**Table 8.1:** Variable factors which influence the costs of irrigation

| **Factors** | **Characteristics** |
|---|---|
| The size of the irrigation project | • Costs of water perform reservoirs tend to be lower for large-scale projects<br>• Studies in Madras, India, showed costs of water, based on the construction of dams, ranging from 6-9 cents per $m^3$ for small dams to 0.7 cents for the largest dams |
| The type of water supply | • Costs of water for irrigation are naturally dependent on the source of water supply, being generally lowest when drawn by gravity from flowing streams or springs, higher when pumped from shallow wells and streams and highest with water pumped from deep wells and reservoirs created by constructing large-scale dams. |

| | |
|---|---|
| Whether the project is government-sponsored or privately-sponsored | • Although theoretically, the cash costs should be the same regardless of who sponsors an irrigation project, government-sponsored projects have been found to be more expensive than private projects, perhaps because only governments are willing to undertake those that are costly in both total and per terms.<br>• It has also been found that private owned water sources offer greater availability and reliability of water supply than State-owned water sources, which is reflected in the higher cropping pattern and cropping intensity in the privately owned farms. |
| The type of crops grown | • The types of crops grown and their market values are important factors which largely influence the profitable use of irrigation water.<br>• Most farmers grow only the most valuable crops at any particular time.<br>• The cropping pattern is much influenced by the availability of good-quality water, assured market, the market price of the crop during that season, enough labor during that season and crop rotation requirements. |
| The climate of the area | • The climate of an area in an irrigation project has a bearing on the irrigation costs and benefits.<br>• In heavy rainfall areas, a few acre-inches of irrigation may be enough to supplement the rainfall whereas in dry areas plant water requirement is met through full irrigation |

**Source:** https://www.greenscapeinc.com/blog/what-factors-affect-the-cost-of-irrigation-services-in-raleigh-durham

Risk and uncertainty are major problems affecting farmers' decision variables. This has prompted irrigation researchers to identify the impact of irrigation on farmers' risk reduction and decision making.

The benefits that accrue from irrigation are:

- Direct incomes resulting from increased crop yields and quality
- Indirect benefits resulting from increased employment, insurance against famine and reducing population pressure

Although the economic and financial aspects of any investment must be considered carefully before a final decision is taken, many Governments in recent years have found themselves compelled to embark, for social and other non-technical reasons, on both large-scale and small-scale irrigation schemes which may not have appeared strictly economic in bankability terms. Irrigated land can support a larger population than unirrigated land. Population absorption and employment generation have been and are likely to remain major objectives of settlement projects in Tropical Africa.

The amount of water for which a farmer can find remunerative use depends on the price of his products and the inputs, and also the cost of obtaining the water, which normally differs from country to country. Even where the economic and financial feasibility of irrigation has been accurately confirmed, there is a need for frequent reevaluation and re-appraisal as changes in prices of the main investment items and farm inputs and of farm produce occur. Superior technology may also improve the yields and quality of the crops and this should be taken into account (Chuet al, 2015).

For economical and efficient distribution of irrigation water, the operator must at all times have complete control of the water as it flows from the head ditch onto the land. This is true, whatever method of irrigation is used. When uncontrolled streams of water are turned into the fields, waste, inefficiency, and uneven distribution are almost certain to result. This can be averted by the use of relatively simple equipment which provides a means of distribution and control of water.

## 8.6 METHODS OF IRRIGATION

Irrigation water is applied to land by three general methods namely:

- Surface application by flooding
- Sub-surface or with furrows in which the surface is wetted little if any
- Sprinkling, in which the soil surface is wetted as much as it is by rainfall.

The surface and sub-surface methods are further subdivided as follows:

- Surface application: Uncontrolled or wild flooding; flooding controlled with corrugations, borders, and basins; and Furrows.
- Sub-surface application: Controlled by lateral supply ditches; uncontrolled irrigation through the excess application of water to adjacent or higher lands

Irrigation methods vary in different parts of the country and even on different farms within a community because of:

- The difference in soil topography
- The difference in the water supply
- The difference in the crops grown
- The difference in the customs

Close-growing crops such as lucerne, clover, and pastures are normally irrigated by use of borders and basins. Forages and some vegetable crops are all suited to flood irrigation by borders and basins. Row crops generally are irrigated by furrows. Any one or a combination of several methods may be best suited to one. Although sprinkler irrigation is used by a few farmers whose farms are very uneven the most widespread method of irrigation is by flooding, especially basin flooding and furrow irrigation. Sandy soils usually have too high an intake of water. In view of this they are best irrigated by over-head means, but this method is far more expensive than flood irrigation. The essential requirements for flood irrigation are sufficiently smooth land of very gentle gradient; preferably flat (Chuet al, 2015).

**Figure 8.1:** A typical irrigation system in arid and semi-arid areas.

**Source:** http://news.mit.edu/2015/quantifying-impacts-of-large-scale-irrigation-on-rainfall-0907

Where flood irrigation is being carried out there must be complete control of the water and the farmer must know the amount of water his layout is capable of applying. Uncontrolled irrigation leads to:

- Overwatering
- Inefficient water-use
- Poor crop returns

Flood irrigation is generally the simplest and cheapest method of applying water. However, it has its own disadvantages in that:

- It requires constant attention as regards maintaining levels and smoothness
- More skilled labor is required to apply water evenly and to avoid waste through excessive runoff than for most other methods
- The canals require constant cleaning and maintenance

The one great advantage of a flood scheme is that it is very flexible. If necessary it can be adjusted easily by enlarging the volume of water to increase the acreage irrigated or to irrigate the lands mare quickly than originally designed.

Row or furrow irrigation is probably the cheapest method of obtaining efficient irrigation. It is suitable only for crops which can be planted in rows sufficiently far apart to allow furrows to be made between the rows. Each row may have a furrow serving it or where the rows are fairly close together, one furrow may serve two rows of plants. Irrigation water is run between the crop rows. The size of the stream is varied according to the gradient and soil texture. For flat gradients, long runs, and sandy soils, a large flow of water. In any case, the ideal stream is of such a size as to run to the end of the furrow with inflow just equaling the infiltration in the furrow (Ouda, 2015). Row irrigation is the most common method used to irrigate the banana crop, for example.

Basin flooding is widely used in irrigating fruit trees and vegetables on flat topography. Basins are flat areas surrounded by low ridges or dikes. They may be square, rectangular, or irregular in shape and may vary in size from 6 feet to an acre depending on the soil texture and the size of the irrigating stream. The more porous and sandy the soil is and the smaller the irrigating stream, the smaller should be the basin for efficient irrigation. Where vegetables are irrigated, each basin is made and leveled independently of the others. The basins are filled with water to a depth of 2 to 6 inches depending on the soil type and the crop. Sandy soils require more water because the infiltration rate is higher. Water from the source is led through the main canal, the size of which depends on the size of flow and the size of the irrigating system.

From the main canal, the water is diverted into an intricate canal system which distributes it throughout the farm. The most common means by which this is done is with open ditches or laterals. The ditches are generally permanent features and commonly follow boundary lines, fences, and the edge of fields. They are frequently earth ditches which may suffer from excessive losses to seepage and evaporation, especially in arid regions or in areas of porous or sandy soils. Leading from the permanent open ditches are the field ditches which may or may not be plowed in at the end of the growing season. Water is delivered through the field ditches by means of check structures or turnouts. They usually consist of metal or wooden fixtures, though they may be merely gaps cut in the ditch bank. From the field ditches, water is led into supply ditches and finally into individual basins one at a time by cutting a gap in the levels surrounding the basins. This method of irrigation does not allow heavy mechanization. It requires much human labor with a high degree of skill for adjusting the flow of water to avoid over-watering or under-watering (Ouda, 2015).

To secure satisfactory returns from irrigation farming;

- The farmers must have knowledge of the important factors affecting returns.
- They must also be proficient managers capable of taking advantage of every opportunity to secure increased returns from their crops.
- For successful irrigation farming, experience in irrigation work is essential.
- Farmers must know when to irrigate. This is important because expensive water may be wasted through over-watering.
- It is essential that the farmer has a high degree of managerial ability and be able to accept new ideas and innovations
- The response in yields due to irrigation is only realized at high levels of complementary inputs such as fertilizers and manures. Without good husbandry practices, money spent on irrigation would be money lost in the drain.
- Successful irrigation also requires that a farmer have first-hand market intelligence so that he can make quick decisions on types of crops to grow, how much of each crop to grow, and in what

seasons to grow them. The aim should be to grow crops that have the highest return per acre-inch of water applied.

- Good irrigation farmers stagger the planting dates of various crops in such a way that the crops mature at different times. This is one of the advantages of using irrigation water. In this way, the farmers get a regular income by maintaining a constant supply of produce.

Final note:

This chapter discussed the systems and importance of good and well planned irrigation systems.

# Man's Response to the Impact of Global Warming

# 9

## CONTENTS

9.1 Introduction .......... 176
9.2 Energy Management .......... 178
9.3 Nuclear Power Energy Application .......... 181
9.4 Solar Energy as an Effective Renewable Energy .......... 185
9.5 Rain Water Harvesting .......... 189
9.6 Agricultural Crop Adaptation to Effects of Global Warming .......... 193
9.7 Insurance and Compensation: Agricultural Products .......... 195
9.8 Afforestation as a Mitigating Effort Towards Global Warming Impacts .......... 198
9.9 Reforming The International Institutions That Deal With Impacts of Global Warming and Climate Change .......... 201
9.10 Key Challenges in Mitigating the Impacts of Global Warming .......... 206

## 9.1 INTRODUCTION

Adaptation to various conditions of global warming is a necessary outcome. This adaptation may further be facilitated by technological advance and control over nature. In societies where technological advance and control over nature is minimal, other social mechanisms may develop which evidence man response to the impact of globalization is warming. Late rains, for example, ruining the harvests and causing famine, may mean unrest and consequent dispersal of a whole community, in search of other alternatives to solve the problem of food shortage. In some cases, it could be permanent. It is here assumed that such movements as seasonal migrations, in addition to reflecting food shortages and other problems that face people in given environments, could also have significant effects. It would interfere with existing social relations and roles. An example would be a mother's absence from home for some time, in the quest for food. Such a mother's absence from home may mean the husband having to play some of his wife's roles (Winston, 2014).

The relationship of the members of a family and the roles allocated to each may be diverted from the normal day to day expectations subsequent upon an ecological problem or requirement. Either the man or the wife might have to change his/her pattern of role expectations. The wife, for example, may have to act, for some time, as the household head, or the man may have to do the duties of the wife for some time. Children's roles may also differ depending on ecological conditions, and it might also hold true, for some social groups, that the household or family authority structure bears and displays some degree of relationship to the particular group's concept of environmental expectations.

The structure of the family may also be affected by long absences of adult males or household heads. This argument may be examined in relation to harsh physical conditions as would be found in the drier areas, where some males may be moving from one area to another, looking after their cattle, especially in dry seasons when pasture shortages are likely to occur. The case of the Gwembe Tonga of Zambezi Valley is a good example of a recurrent hunger area with low rainfalls, erratic floods and insect pests being among the factors which in any year might bring about hunger to some neighborhoods or at times to all of them. In one of such occurrences, this led to periodical labor migrations of the adult males and during their absence, their wives cultivated the fields and provided for the family subsistence (Winston, 2014).

Components of the physical environment are broken down to such characteristics as:

- Rainfall
- Altitude
- Temperatures
- Soil fertility
- Vegetation
- Ecosystem

Adequate availability of such factors as rainfall and fertile soils may mean an abundance of agricultural harvests, including cash crops, while the reverse such as shortage of the same factors may spell the occurrence of such phenomena as droughts and famines, to mention two of the many possible consequences of global warming.

These characteristics of the physical environment then may, either in a combined or singular form, influence economic modes of livelihood such as livestock keeping, where livestock keeping in particular ecological settings is one of the most optimal land utilization alternatives, given the scarcity of such resources as rainfall, pasture, fertile soils, etc.; or other agricultural practices and types of crop as may be suited to various kinds of physical environmental qualities as outlined above. There are certain social and sacred activities or ceremonies that do prevail at certain times of the year. In such cases, such activities could be reinforcing some roles while de-emphasizing others at the same time. In addition, the activities referred to may be supporting some belief systems or to the contrary some belief systems may develop to justify the indulgence in some particular activities, or as psychological mechanisms to support a meaningful and satisfactory existence under certain conditions (Winston, 2014).

In cases of regularly anticipated environmental problems, the structuring of the societies may develop responsive mechanisms toward their environment. Such behavior, adapted to the natural setting may be reflected, for example, in kinship, neighborhood and hospitality patterns and even territorial complementarity of scarce resources. The traditional social insurance systems such as the famed nomads' reciprocity, communal rights to limited resources such as watering places have all evolved to sustain the survival of both the social and economic systems in such adverse environments. The social status or ranking criteria for such nomadic groups may differ considerably from those of the rich highland agriculturalists.

Societies in different environments would need to be evaluated in their own settings, if possible, to find out to what extent their social structures are molded by such environments and whether generalizations, involving two or more societies can safely be made without running into serious difficulties among which could be wrongly biased statements and fallacies if, on such understanding, generalizations are not possible, then common yardsticks for, say, what type of projects and/or techniques are needed for development, social or economic, to benefit such groups are not commonly applicable. Awareness, therefore, of the impacts of global warming and other climate change on social structures, leading to increased understanding of environmental stresses' and their impingement on people's social and economic organization would facilitate the characterization of societies into more helpful categories. Such categories can then be used as developmental targets and this might help overcome some of the problems arising due to the placing of all societies or a group of them under general or broad categories with the assumption that a given developmental strategy will apply equally to all of them (Winston, 2014).

## 9.2 ENERGY MANAGEMENT

There has been an increased concern over the energy consumptions in the world due to the impacts of global warming and climate change. Different energy sources have been being used over the past decades. Fossil fuels have been challenged by pollution as well as increased impacts such as global warming and climatic changes. The demand for energy has been increasing; however, the availability of energy has become costly due to depletion of the non-renewable sources. The alternative energies such as renewable energy sources such as solar energy and wind energy have been proposed. The building sector has been affected by increased energy demand. The net-zero energy building (NZEB) has been introduced as the new architectural designs for the building (XinRong and Ibrahim, 2016). This technology is a building design whereby the amount of energy consumed annually equals the amount of onsite produced energy. The energy management in the net zero energy building involves the use of:

- The grid connections
- The renewable energy certificates
- Building environment niche
- The net-zero energy building technology

Some countries have embraced the net zero energy building technology as a building and construction technology. For instance, the European states have formed the energy commissions that regulate the compliance of net-zero energy building requirements by member states. This technology is a relevant tool for reducing energy consumption as well as alleviating the impacts of global warming.

Environmentalists all over the world have raised concern over global warming as a result of the greenhouse effect. The buildings also contribute to the amount of carbon dioxide emitted into the atmosphere from the combustion of fossil fuels. The net zero energy building uses the green energy approach; the emission-free energy such as wind energy and solar are used in the provision of energy (XinRong and Ibrahim, 2016). The energy is produced within the net zero energy building hence the costs of transmission are limited. This technology is a significant strategy in a bid to lower the pollution as a result of the carbon dioxide emission.

- The net zero energy building is a multifaceted value creation avenue. These technologies are expected to revolutionize the competitive advantages in the architecture as people approach towards the future of building technology.
- The value of properties such as land, industrial plant or residential building would improve with embracement of the technology.
- The market risks are supposed to be mitigated with the introduction of the net zero energy buildings. For instance, the increased energy costs will be reduced when net zero energy buildings are used.
- Since the technology utilizes safe energy, the welfare and health of the inhabitants are promoted. For instance, respiratory diseases from the smokes are eliminated in the net zero energy buildings.
- The technology saves money; there is about 60% to 90% energy efficiency.
- Using the net zero energy building architectural designs saves money over the whole life cycle of buildings; the energy maintenance costs are brought down.
- The application of technology is educational. The people who participate in design, construction and maintenance of a net zero energy building gain important knowledge in the energy management and energy efficiencies.

Although the concept of the net zero energy building seeks to reduce the average net energy per annum, it faces some drawbacks.

- There are times when the net zero energy building will require the support of the grid connection to provide energy.
- The renewable energy sources have reliability problems; there are fluctuations in the capacity of renewable energy sources to produce adequate energy. When renewable energy sources become unreliable, the net zero energy building depends on the electricity grid whereby the efficiency of its components is compromised.
- Moreover, the efficiency of the net zero energy building depends on the behavior of the inhabitants.
- The energy management strategies depend on the energy consumption behaviors of the inhabitants.
- There are no hard set rules as to how energy is supposed to be used.
- There is a need to embrace the ways that can reduce the impact of these drawbacks.

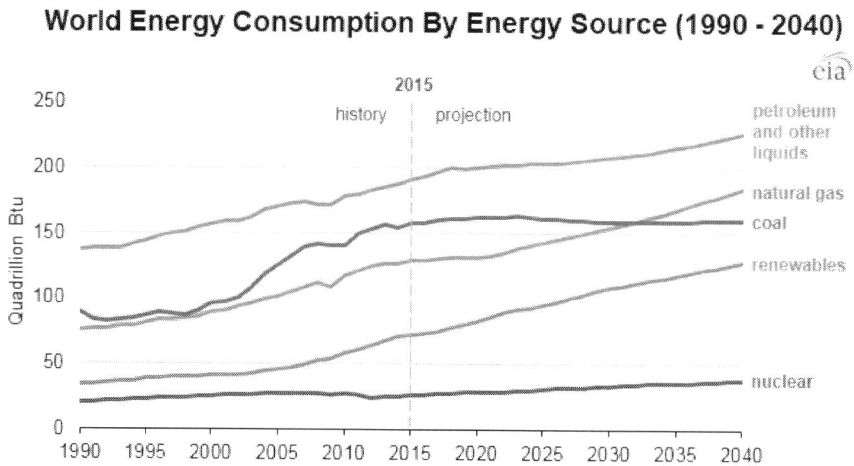


**Figure 9.1:** The global consumption of energy in the world has been increasing year after year.

**Source:** https://www.ecademyonline.com/what-is-the-distribution-of-world-energy-consumption/

The global energy consumption increases from 524 quadrillions Btu (2010) to a projected 630 quadrillion Btu (2020). Further projections show that in 2040, the global energy consumption will have hit 800 quadrillions Btu. The increase translates to a 56% rise in every 30 years and demands for a strategic reduction in energy consumption.

Fossil fuels contributed to most of the global energy consumption. Fossil fuels are one of the most harmful sources of energy. The global warming and consequent climatic changes are driven by the carbon dioxide emission related to fossil fuels. The renewable energy sources contributed to 11% of the global energy source in 2013. However, fossil fuels contribute to about 90% of the global energy sources. Because fossil fuels are responsible for the increase in the global surface temperatures as a result of the greenhouse effect; there is a need to lower their use. The domestic energy consumption totaled to about 40%. Out of the 40% energy consumption, the fossil fuels contributed to about 30%; meaning that the emission of gases and the subsequent greenhouse effect was high (Shrader, 2011).

The energy consumption from electricity in a typical building is about 70%. About 35% of the electricity used in the building is generated far away and then transmitted using wires. To curb the problems of high energy demand and transmission, the proposed future building will be designed to offer high energy efficiency. Moreover, the integration of renewable sources will promote the energy generation in the net zero energy building. The technology targets to save about 40% energy because of the energy efficient design as well as utilization of renewable energy sources. The proposed net zero energy building design offers the production of energy that is later stored in the ESS (Energy Storage Systems).

## 9.3 NUCLEAR POWER ENERGY APPLICATION

Despite the negative uses of nuclear energy that engulfed a score of quarters, where some still believing that it could be used again like in 1945 in Hiroshima and Nagasaki by the US, nuclear power has been championed to successes the previous sources of energy that has detrimental effects on the environment and biodiversity. To prove their commitment to ensuring that the nuclear power energy production and usage comply with the prescribed standard, different bodies have been set by different countries where nuclear energy is commercialized (Shrader, 2011). Such bodies include:

- Canadian Atomic Energy Commission
- The Southern Asian Nations (ASEAN)

- The United States and European countries
- International Atomic Energy Agency (IAEA)

The initial step toward ensuring that nuclear power is a safe source of energy was initiated on 12 September 1945 by Henry Stimson, a German Secretary of War in a recommendation he was making to his president, Truman. In his view, the USA was to agree with former USSR to develop and use atomic power for purposes of improving the lives of humankind. To date, the spirit of safety has been lit to higher heights with more countries joining the bandwagon. For instance, to demonstrate the safety of nuclear power as a source of energy, the World Nuclear Association reports that there have been only three major accidents reported for nuclear reactors since civil nuclear power was initiated. This is with cheer recognition that there are over 14,500 cumulative nuclear reactors committed to full-time commercial services of nuclear energy production in 32 countries (Shrader, 2011).

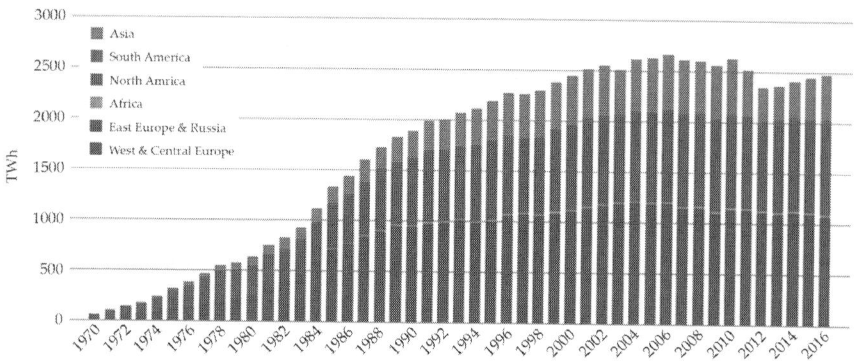


**Figure 9.2:** The trend in nuclear power production by country.

**Source**:http://www.world-nuclear.org/information-library/current-and-future-generation/nuclear-power-in-the-world-today.aspx

The success story is pegged on the fact that the security measures in nuclear power plants are always robust with no single loophole left for any mishap. According to the same report, the nuclear power plants sites are also among the safest places to work. Apart from the tragic Chernobyl accident in Ukraine 1986, there has been no report of any worker or citizens who have ever lost their lives as a result of exposure from radiations emanating from commercial nuclear plants. In addition, the association categorically reports that nuclear power reactors dedicated to commercial services cannot explode any reported circumstances since the fuel used in them is never enriched past about 5 percent. Such prescriptions are monitored by IAEA which is mandated with the role of prescribing safety procedures and reporting any minor irregularities in compliance.

Noting the magnitude of potential damage that can result from improperly handled nuclear power, its acceptance as a safe source of energy is vested on a wide range of factors that are piously hindered. The commitment of the various institutions that oversee the implementation of the laid down standards leaves no doubt that nuclear power energy is certainly a safe source. In addition, the following factors may also shed some light on the proposition of why nuclear power is deemed a safe source (Shrader-Frechette, 2011).

### 9.3.1 Licensing

It is a legal requirement as per the UN and IAEA standards that all nuclear power plants and reactors be licensed. As International Nuclear Societies Council (1992) observed during a conference in Rio de Janeiro, licensing of nuclear plant have compelled the management to only utilize qualified personnel; hence, reduce any chances of unwarranted mistakes. In addition, the licensing standard compels that all nuclear power plants should have a pressure-resistant containment building. Therefore, unlike other alternative sources of energy, nuclear plants have the highest levels of amassed safety measures. In addition, any new design must comply with some of the stringent safety standards that are inspected by IAEA and the International Nuclear Societies Council before any activity commences. These bodies are concerned even with basic things like the location of the plant. If it does not comply with the standards, such a site is ruled as not fit; hence, construction cannot proceed. This information is captured on the Nuclear Power Licensing Reform Act of 2011 that amended Section 103 of the Atomic Energy Act of 1954 to include such licensing requirement (Andrews, 2012).

### 9.3.2 Environmental Impact

There is some score of individuals who still hold that nuclear power is not a safe source of energy; hence should be discouraged. On the contrary, nuclear power is among the safest source of energy power. The only reported meltdowns that warrant cautions are where human errors are pronounced. The Fukushima and Chernobyl accidents were purely as a result of human errors, and the same is bound to happen even in other alternative sources. In regard to environmental impact, nuclear power has no known waste emission or radiation, unless there is leakage. In addition, unlike coal, oil and other natural gases that run out with time, nuclear power is efficient, cost-effective and their sources do not wear out that fast.

The safety of these power sources is also guaranteed by the presence of the new high-temperature reactors that are perfect in ensuring that sea waters are not polluted. Burning such products like fossil fuels, it is said that approximately twenty billion tons of ozone harming gases like carbon dioxide are released in the environment. This is not the case with nuclear power energy that has very little or no ozone gas that is injected in the environment. It is worth noting that the little nuclear wastes that are produced in power plants are deposited in deep geological storage sites where they spontaneously decay with time. The little amount is also enabled by the fact only a very little amount of uranium is required to produce energy equivalent to the one produced by burning tons of fossil fuels (Andrews, 2012).

### 9.3.3 Technology

In the height of so many controversies surrounding nuclear power energy production and usage, very stringent precautionary measures have been employed to guarantee safety. The measures and standards are particularly enabled due to advanced technologies that have been employed in the power plants. Technology has enabled effective use of natural Uranium in such a way that the nuclear wastes produced are also exploited to produce more energy. The whole process takes place in a highly designed reactors called reactor breed. Technology has also enabled continuous improvements in the creation of new reactors design which include extremely advanced safety features. Like in the reactor breeder, recycling and reuse of nuclear fuels that has been facilitated by technology. Technology, as it was noted during an international Ministerial Conference on Nuclear Power in the 21st century held in St Petersburg, has provided an opportunity to supplement nuclear energy with thorium as an alternative fuel (Shrader-Frechette, 2011).

Use of technology has also facilitated for economical, clean and safe nuclear energy even in the developing countries. For instance, due to advancement in technology, the western world is now able to operate its nuclear power plants using what they call as defense-in-depth approach which entails application of multiple safety systems that complement the natural features of the plant's core. Such technology allows for comprehensive and timely monitoring and progressive testing so as identifying any chance of equipment or operators failures. Technology has also allowed for provision to confine the aftermath of unwarranted fuel damage or other technical problems to the reactors. Employing technology has also facilitated the usage of equipment s that has the potential to prevent operational disturbances or human failures and errors exploding into unmanageable problems (Shrader-Frechette, 2011).

Carefully noting the assertions that are meant to discourage nuclear power usage, regard it as unsafe, it is worth retaliating that statement is far from the truth. The fact of the matter is, nuclear power is a safe and reliable source of energy. It has the potential to solve some of the fears of reduced power sources in the future. Nuclear power energy also is better placed to curb the rising instances of environmental pollution arising from alternative power sources. With the commitment demonstrated by countries already using nuclear power as a major source of its energy, it is clear that even the slightest doubt of this source being unsafe is removed. In fact, adopting this source will not only mean that the world is safe from environmental pollution, but also the challenges of insufficiency, depletion and safety are also solved (Shrader-Frechette, 2011). Therefore, nuclear power as a source of energy should be deemed as a clean and safe source but not otherwise.

## 9.4 SOLAR ENERGY AS AN EFFECTIVE RENEWABLE ENERGY

There are a number of good reasons in favor of the development and widespread use of solar energy resource in the world. First, there is an abundant supply of solar energy. Solar energy also provides a truly renewable resource. Its use would open up an indigenous, relatively secure energy option. Furthermore, its use would make possible the release of other fuels for premium use. Studies have shown that a large part of the population lives in rural areas, the cost of providing reliable fuel transport and/or electric power transmission networks is excessive and as such, localized solar energy installations are cost effective. It is also important to note that solar energy technology is relatively unsophisticated compared to, say, nuclear engineering. Ultimately, solar energy is non-polluting (Spellman, 2014).

**Figure 9.3:** Solar power application.

**Source:** https://www.pinterest.com/pin/825777281649439041/?lp=true

There is a need to reduce the proportion of fossil fuel products in the total energy mix. With the expected growth in the world's economy, the total energy consumption is bound to increase. One way of reducing the consumption of imported fuels is to institute energy conservation measures. This would reduce the total energy consumption by improving the efficiency of utilization of the existing energy supplies. Continued development and utilization of renewable energy resources available locally would certainly go a long way to reducing the dependence on imported fossil fuel products. The utilization of solar energy (and other renewable energy resources) will thus contribute significantly to alleviating over-dependent on fossil fuel products, thus reducing the impacts of global warming and climate change.

Above the earth's atmosphere, the flux of solar radiation is approximately 1400 W $m^{-2}$. However, by the time it reaches the earth's surface it has been reduced to between 1050 and 0 W $m^{-2}$. This reduction is due to reflection back into outer space and absorption or scattering by molecules; this is especially so with the scattering by water droplets in the clouds and by dust particles. The intensity of the radiation reaching the earth's surface at any one time thus varies from place to place. In addition to this geographical variation, there is also a seasonal and diurnal variation of solar radiation intensity at a particular place. Reliable solar radiation data is required for the design and prediction of the long-term performance of any solar energy system (Spellman, 2014).

The most common instrument for measuring the duration of sunshine is the Campbell-Stokes recorder. It consists of a glass sphere which focuses the solar rays on to a sensitized paper, thereby causing a burn to appear on the paper whenever the heat rays are powerful enough. The times during which burning occurs are taken as periods of "bright sunshine". Other sunshine recorders exist. The main one, used in the USA, consists of mercury in a tube. The mercury expands in response to the heat from solar radiation- and thus causes an electrical switch to close. Modified cameras which use photographic paper to record the sun's light rays also exist (Spellman, 2014).

Solar energy can be transformed into heat energy to heat, for instance, air or water which can then be used in various ways. Solar radiation is absorbed by a suitable material having a black surface and converted into heat. The temperature of the absorbing material thus rises. By bringing a working fluid such as water or air into thermal contact with the solar energy absorber, the former extract heat from the latter with a corresponding rise in temperature. The devices which transform solar radiation into heat energy in the manner described above are called "solar collectors". There are two main types of solar collectors: flat-plate collectors and focusing collectors.

However, one of the major problems to be overcome in utilizing solar energy in continuous processes is its intermittent nature. For instance, solar energy is not available in sufficient quantity during cloudy spells and at night. Some degree of storage is therefore required in any application of solar energy. In thermal applications of solar energy, the problem becomes that of heat storage. The simplest way to store thermal energy is as sensible heat, heating a suitable material to a higher temperature. Usually, water or rocks are used as storage media. The disadvantage of sensible heat stores is that they tend to be large and heavy and containment costs can be high. The alternative system, latent heat storage, makes use of the fact that certain substances undergoing a phase change absorb large amounts of heat energy which they release when the reverse phase change occurs (Spellman, 2014). Therefore, salt hydrates are ideal materials for latent heat storage.

Different researchers have investigated both the technical and economic feasibility of utilizing solar energy for heating in the food processing industry. The applications considered are outlined in the table below.

**Table 9.1:** Utilization of solar energy in the food processing industry

| Type of processing | Characteristics |
|---|---|
| Food drying | • Solar energy has been in used to dry food products such as cereals, pulses, meat, fish, fruits and vegetables from time immemorial.<br>• The food products are simply spread on the ground or on platforms and turned regularly until drying is complete.<br>• The drying period is long and depends on the climatic conditions.<br>• The quality of the dried product is dependent on the climatic and environmental conditions, for example (a) drying is adversely affected by rain and storm; (b) pollution and dust adversely affect the quality of the dried product.<br>• It is difficult to maintain a dried product of standard quality.<br>• The product being dried is open to infestation by insects and intrusion by animals.<br>• Coffee-beans are either sun-dried or mechanically dried. Mechanical drying involves passing heated air through the coffee beans.<br>• In tea, drying is achieved by passing hot air through the fermented tea.<br>• In fish processing, sun-drying is the simplest and most economical method of fish preservation.<br>• Sun-drying fruits and vegetables is an ancient method and it is still in use in many parts of the world. |
| Solar Thermal Systems | • Solar energy could be used in a variety of industries where the generation of hot water (or low-pressure steam) is required.<br>• Hot water is extensively used in the food processing industry for a variety of applications.<br>• The most prevalent of these are washing and rinsing.<br>• The temperature of water for cleaning and rinsing in food processing plants varies from 38 to 82°C.<br>• There is certainly a potential for the use of solar energy systems in the world as most parts of the world receive large amounts of solar energy throughout the year. |

| | |
|---|---|
| Refrigeration | • Perishable foods such as fruits and vegetables, milk, fish and meat require refrigeration facilities to extend their shelf-life.<br>• Adequate refrigeration facilities will facilitate the transportation and marketing of these products.<br>• Solar energy can be used to operate certain refrigeration systems.<br>• The use of solar energy to power refrigerators would reduce the operating costs of refrigeration systems.<br>• Moreover, solar-powered refrigerators can be used even in remote rural areas without electricity supply.<br>• The use of solar energy for refrigeration is attractive because refrigeration is most needed during periods of high solar radiation intensity. |

**Source:** http://ktisis.cut.ac.cy/bitstream/10488/880/3/C44-IPH-NREC.pdf.

## 9.5 RAIN WATER HARVESTING

Most places in the world have become arid and semi-arid due to the impact of global warming and climate change. These places are characterized by:

- Erratic rainfall with great annual variability about the mean
- Annual potential evapotranspiration exceeding the rainfall
- High amounts of runoff due to low infiltration into the crusting soils
- Recurrent soil moisture deficits limiting crop production

Rainfall is the main source of water in dryland farming. In arid and semi-arid areas, rainfall varies considerably in time. Crops are adversely affected by water stress depending on the crop adaptability to the environment. Agricultural crop production is always a problem in these places because of the lack of adequate soil moisture. The aim of a water harvesting system is to supplement rainwater with runoff water and alleviate possible soil moisture deficits in critical crop growth stages. This situation can be mitigated through two practices, which are irrigation development and rainwater harvesting (Celesteet al, 2014).

During high rainfall, runoffs can always be harvested and stored for agricultural purposes. For instance, a country like Kenya receives bimodal rainfall. These rains are normally received on March (long rains) and October (short rains). In both seasons there are wide variations in the distribution of rainfall. The rainfall is often of very high intensity and short duration hence generating very high amounts of runoff. Up to 64% of rainfall may be lost as runoff from degraded surfaces (Celeste et al, 2014).

The high runoff rates are attributed to the low infiltrability of the soils due to their surface sealing and crusting properties. To sustain crop growth during critical crop growth periods, it is advisable to always examine methods of conserving enough soil moisture.

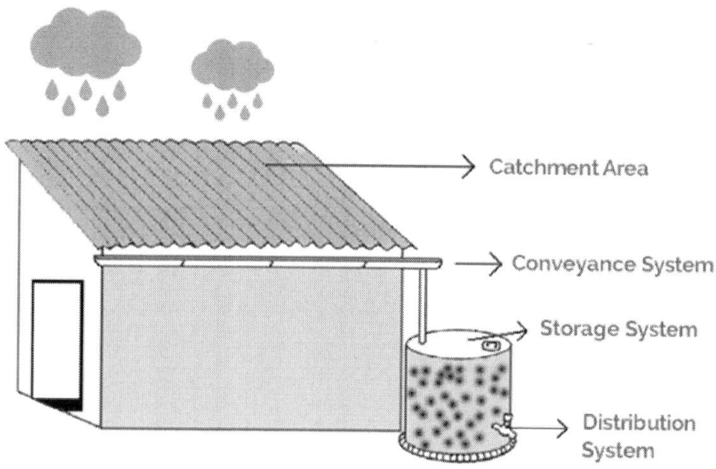


**Figure 9.4:** A typical rainwater harvesting system.

**Source:** http://www.neoakruthi.com/blog/rainwater-harvesting-for-house.html

Many dry areas, especially in Africa receive rain for up to 50 days after which infrequent rains adversely affect crop yields of the rain-fed crop. To reduce the resulting soil moisture shortage, the magnitude of moisture deficit and the time it is likely to occur within a crop growing season must be known.

The probability of a soil moisture deficit occurring within the growing season should be studied using years for which complete daily rainfall data over both seasons are available. The method involves identifying the onset of each "long rains" and "short rains" seasons, which come in March-May and October-January, respectively. The onset should be taken as time after the amount of rainfall received in one day or total for 2-3 consecutive days. After this onset, rainfall should be accumulated for the next 90 days. A mean of the totals should be obtained. The data is then examined to separate the "wet" from the "dry" years. A wet year should be taken to have a total above

300 mm "long rains" and above 450 mm "short rains" which is the 24-year mean and dry year a total below these figures for the seasons. A daily average of the "wet" and one for the "dry" should be computed to represent the average growing season (Celesteet al, 2014).

Numerous researchers have considered rainwater harvesting as the most promising agricultural system for arid and semi-arid places. Water harvesting systems have been in use for in different places for a long period of time. For instance:

- For twelve centuries In the Negev Desert, peaking around AD 100 to 700 when 4000 ha were cultivated using this technology
- The technology has been in use for some 9000 years in Jordan
- The Marib diversion dam was used around 750 BC to raise the water level for irrigating adjacent fields in Yemen
- The technology has also been used in the Thar Desert (India) and in the USSR
- In North Africa, complex systems have been used in Egypt, Libya, and Algeria
- 10 million olive trees are still sustained by harvested water in Tunisia

It is vital that runoff water is harnessed and conserved in situ. Due to the significant loss of the limited rainfall as runoff in arid and semi-arid areas, there is need to evaluate potential runoff water harvesting techniques that would provide adequate soil moisture to sustain crop growth during the dry spell. The in situ conservation of runoff requires some comprehensive understanding of soil moisture dynamics as influenced by soil properties and rainfall characteristics.

The amount of soil moisture available for plant growth would depend on the amount of runoff generated, and the amount subsequently conserved in the soil. The rates at which the impounded water infiltrates into the soil or is lost through evaporation or deep percolation significantly influence the soil water balance (Celesteet al, 2014).

The traditional systems used by farmers in arid and semi-arid areas to grow crops are well -adapted water harvesting systems. The most commonly used harvesting structures are:

- Rock catchments
- Harvesting from roofs

- Semi-circular hoops
- Trapezoidal bunds
- Terraces (are widely used for moisture conservation)

Agricultural crops planted using water harvesting normally have a 95% survival rate, and yields improve significantly.

**Figure 9.5:** In situ rainwater harvesting technique.

**Source:** https://www.researchgate.net/figure/Overview-of-soil-and-water-conservation-measures-in-the-steep-sloped-area-for-reducing_fig8_288477784

The heart of a water harvesting research is the water balance of the crop root zone. It is, therefore, recommended that more work is done on runoff rates, drainage, and evapotranspiration. On drainage, such work would be directed at verifying and calibrating the field capacity (FC) model for determining the percolation rate from the crop root zone by using easily measured parameters such as:

- Hydraulic conductivity
- Soil porosity
- Soil moisture status

Alongside such studies, the complex rainfall-runoff relations should be looked into for several years. This would reveal the influence of slope, soil type, vegetation and antecedent soil moisture on runoff. Additional studies are also recommended for local calibration and testing of the runoff model for high intensity, low duration tropical storms incident on arid and semi-arid soils, particularly catchments predominated by soils with surface crusting properties (Celesteet al, 2014).

## 9.6 AGRICULTURAL CROP ADAPTATION TO EFFECTS OF GLOBAL WARMING

Different parts of the world have different climatic because of the impacts of global warming. Many governments have taken the initiatives to develop agricultural production (both crops and livestock) of the arid and semi-arid areas, the production of drought-resistant crops as a way to alleviate the effects of global warming and climate change.

Different cultivars have been developed and are being evaluated in different parts of the world. In any crop evaluation program, the ability of a genotype to perform well over a wide range of environmental conditions is always appreciated. This general adaptability is of significance because of edaphic variation between locations and seasons in any one locality. Crop scientists consider stability in performance as one of the most desirable properties of a genotype to be released as a variety for wide cultivation. No breeding work would be considered complete until the improved genotypes are tested over a wide range of environmental conditions involving different locations and different seasons (Shyam et al, 2011).

There are several statistical methods used for estimation of phenotypic stability. Some of these methods include:

### 9.6.1 Multi-locational Trials

Data obtained is subjected to environment-wise (location) analysis of variance followed by pooled analysis of the data. This has been done in different countries for various crop cultivars. The cultivars which are used in such studies are improved further. In case of the genotype - environment interaction is found to be significant; one of the various approaches known for measuring the stability of genotypes can be used and the genotypes may be ranked accordingly.

### 9.6.2 Regression Analysis

In this technique genotype, - environment interaction is partitioned by calculating a regression of the yields of a given genotype in different environments on the respective means of all genotypes. Phenotypic stability factor (SF): It takes into consideration the mean values only in the highest and lowest yielding environments. This method has been used only to a limited extent because the SF has some very obvious defects as it does not consider intermediate environments thus introducing some bias.

### 9.6.3 A Dynamic Approach to the Interpretation of Varietal Adaptation

It has been showed that regression of yields of separate genotypes accounts for a large part of genotype x environment interactions. This leads to the realization that the components of a genotype and the environmental interactions are linearly related to environmental effects when these effects are measured on the same scale as the genotypic effects. The observed values are regressed on environmental indices defined as the difference between the marginal mean of the environments and the overall mean. The regression coefficient is then taken as the stability measure. It has been shown that the phenotypic stability of the varieties is inversely proportional to the mean yield. From the analysis, the varieties specifically adapted to good and poor environments and those showing general adaptability can be identified. Varieties from particular geographical regions show a similarity in the type of adaptation which provides a useful basis for plant introduction (Shyam et al, 2011).

Most of the crop breeders use regression analysis techniques. This is because, although genetic effects are not generally independent of environmental effects many authors have observed that the relation between the performance of different genotypes in various environments and some measure of these environments is often linear or nearly so. It has been suggested that fundamental statistical assumptions are not usually satisfied and that the choice of the sums of squares and degrees of freedom from which the regression components are subtracted is not appropriate. Secondly, the use of the mean yield under each environment as the environmental value of the particular set of genotypes as an independent variable which is in fact not independent of the phenotypic variable regressed onto it has been criticized.

Shyam et al, (2011) discussed the significance of the genotype - environment interaction on the basis of the relative magnitude of different variances estimated from multi-location-year testing. They suggested that heterozygous and heterogeneous populations offer the best opportunity to produce varieties which show small genotype x environment interactions.

When developing crop cultivars, breeding is repeatedly done at one location and testing of the improved cultivar may be done at several other locations. A high yielding genotype in a favorable environment does not mean that the same genotype will also be high yielding genotype in an adverse environment. The reason for this is that although it is the same genotype, the combination of different genes are called into play in the two different environments hence making this genotype behave as if it were two different genotypes in the two environments.

## 9.7 INSURANCE AND COMPENSATION: AGRICULTURAL PRODUCTS

People always behave rationally to maximize gain. In other words, this means that men behave according to anticipated rewards, and where faced with competing choices, they will choose the option which carries higher rewards. The rewards could be material things or they could be purely social and psychological like acceptance, prestige, sympathy, praise or esteem. As in all situations of social interaction, ideology (shared values) beliefs, goals and expectations are the motivating factors. Motivation to act derives out of the probability that the interacting individual's goals will be realized or interests served. In livestock farming, insurance is an important social security system. It is established and maintained by the constant exchange of livestock gifts. Livestock farmers should procure insurance for their stock to enable them to reciprocal partnerships through the distribution of livestock gifts (Olivier and Stutley, 2010).

**Figure 9.6:** Some of the major agricultural products to insure.

**Source:** http://pcic.gov.ph/

### 9.7.1 Definition of the Term Insurance

There is no single definition of insurance. This is because the purposes for which the definitions of insurance are invoked differ.

The following are some of the definitions of the term insurance by different scholars.

- A lawyer: "insurance is a contract whereby one person called the insurer, agrees for a consideration called the premium, to pay to another person called the insured, a sum of money or its equivalent on the happening of a specified event."
- An economist "insurance is a device for the transfer of some risks of economic loss from the insured who otherwise would have borne the risk, to an insurer in return for a premium."
- A Sociologist "insurance is a device whereby the participants provide financial compensation to those among them encountering the many misfortunes or contingencies that befall humanity."

Agricultural practices are full of risks and there is no any farmer is faced with the possibility that one day, those hazards that form part of life will befall him and cause him loss some or all of his agricultural products. The purpose of insurance is to compensate the victim for the losses incurred as a result of a natural calamity such as drought or famine. The concept of insurance is therefore not completely alien to customary or traditional practices. It should be agreed; however, modern insurance as practiced today is foreign in many concepts to the customary practices of many people around the world (Olivier and Stutley, 2010).

### 9.7.2 What is required when procuring a livestock insurance

With livestock, an insurance quote is a basic requirement. Meeting this requirement comes with some financial consequences. Besides cost, one of the best things to do is to have just the right insurance policy. With continuous global warming and climate change in the world, there has been an influx of investors in the livestock industry as well. Some of them are insurance providers who offer completely different policies at different rates. Unfortunately, calamities and disasters are now the order of the day, it is very important to have an insurance quote that addresses your concerns now and in the event of a disaster as well.

Well, with varying terms, it is important to have an understanding of the essentials of livestock insurance policies. With some bit of financial and legal background, you can always get a policy that wholly addresses your concerns. On the other hand, when doubtful or without a concrete background, you can always visit insurance agents. As professionals in this field, they will always be willing to share a good share of what they have gathered over the years. Before procuring insurance for your livestock, it is recommended to always shop and talk to different agents. By the end of the day, you will be able to get a quote that fits your preferences (Olivier and Stutley, 2010).

Getting a good auto insurance quote demands a thorough understanding. It is after this that you can be able to compare quotes from different companies. To many, this step is not always necessary. It is however very important, particularly for a first-time buyer. A good choice is always evident when disaster strikes. The other good thing is that you will be able to save a fair portion of your hard-earned cash. Besides cost, one should always strive to compare the policies, coverage and the benefits that come with different quotes. Read between the lines and understand these key areas. When in doubt, ask to get answers. It is now clear that when searching for the right auto insurance quote, there is more than just the price. Go for the policy that addresses your concerns. On the other hand, you also have a part to play (Olivier and Stutley, 2010).

### 9.7.3 Compensation on Lost and Damage Agricultural Products

The word "compensate" generally means recompense, counterbalance to make amends. The word is derived from a Latin root "*compesare*" which means weigh together and could be expressed as give.to each man that which is due. The principle underlying the award of agricultural loss in terms of crops and livestock due to impacts of global warming and climate change is expressed in the maxim "*Restitutio in intergrum*". This means that the law will endeavor in so far as money can do it to place the agricultural loss in the same position would have occupied had not the loss occurred.

The award of agricultural loss involves the taking away of money from one person and giving it to the other party that has incurred the loss. Before one becomes entitled to receive money in terms of compensation from the other party, he must first prove that the loss has occurred, in this case, death of animals and destruction of crops due to climatic condition. If the loss cannot be proved then the claiming party will receive no relief (Olivier and Stutley, 2010).

Awarding damages and loss for agricultural crops and animals could be regarded as one way of doing this. By far the most convincing reason justifying the award of damages and loss of agricultural products is that of compensation. It is an attempt to make good that which has been lost. It is awarded recognition that damage has been received and that such damage or loss has resulted in a great loss. In assessing agricultural damages, the aim is to consider in pecuniary terms that which has been lost as a result of natural calamities. Under pecuniary agricultural loss, damages awarded include that of loss of future earnings, irrigation expenses, and expenses on chemicals. Pecuniary agricultural loss is awarded for loss of earnings both up to the time of the trial as well as future loss of earnings and for additional expenses incurred as a result of the calamity such as irrigation expenses. Damages awarded should be the exact amount of money lost, but this is only possible with loss of earnings or expenses incurred up to the time of the trial (Olivier and Stutley, 2010).

As far as the prospective loss is concerned its calculation is left to the discretion of the court in exercise. An allowance is also made for the fact that it is being awarded in a lump whereas under ordinary circumstances the claimer would have received it in bits over a long period of time.

## 9.8 AFFORESTATION AS A MITIGATING EFFORT TOWARDS GLOBAL WARMING IMPACTS

Different countries face unique problems with regard to the effect of global warming and deforestation. Some have little or no forests at all. In most countries where some forest exists over-exploitation is a common phenomenon. However, most countries, especially in developing regions, have responded by planting more exotic species with hope of alleviating the impacts of global warming.

The forest sector is poorly developed in spite of the high substantial natural and human resources and the growing demand for forest products in most developing countries. Its low contribution to lowering the said effects of global warming is a resultant of the low priority accorded to it. This reduces the narrow view of forest resources use invalid. Therefore, proper management of natural and plantation forests can on the aggregate contribute to the reduction of global warming. The exploitation of forests is not seen in isolation. It is viewed in relation to other land uses. In most developing countries, governments have commercially feasible forest potentials (García et al, 2017). But their present and future forest cover depends on their capability of forests to compete with other land uses. Some of the practices that compete with land use include:

- Agriculture
- Land for grazing

- Settlements

For a long time, there has been a clash between man and his environment manifested in the conflict between land uses. This is a process which has reduced the protective role of forests in most countries. This is attributed to the high degree of specialization and sophistication among officers working in the field. In international circles, the subject of forest management has been a center of debate for a long time. Much of the discussion is centered on the capability of the sector to alleviate the impacts of global warming and climate change. Most of the problems facing the sector are related to:

- Poor forest covers
- Deforestation
- Poor structure of forest management

Afforestation has been viewed in terms of the need to alleviate the impacts of global warming both locally and internationally. Proposals put forward have been aimed at improving and increasing tree planting activities, from both indigenous and plantation forests. Effective achievement of these proposals demands active participation from the local community. This creates awareness among the people on the need and role of forests in combating the current impacts of global warming and climate change.

**Figure 9.7:** Afforestation as a way of alleviating the impacts of global warming.

**Source:** https://en.vietnamplus.vn/cold-weather-blamed-for-hampering-afforestation-effort/89508.vnp

In most countries, not much information on afforestation and forest management is available. However, it is worth noting that much emphasis is laid on the need to plant more and more trees and manage those trees until they are mature enough. To a large extent forest management in most countries around the world is related to the future utilization of the resource. The issue is centered on consumption versus the supply pattern of timber. It is suggested that the supply will outstrip the demand by five times in the next few decades. Eventually, this will result in extensive forest destruction such that, the presently available resources cannot effectively deal with the impact of global warming (García et al, 2017). Therefore, it is important to conserve and plan for the utilization of forests so that they continue to provide benefits, including alleviating global warming.

One of the major problems attributed to forests has been that of the decline of these resources. This is attributed to the absence of an integrated management plan. In most countries, the department of forests is charged with the responsibility of maintaining a sustained forest cover in respective countries. Over time, the productive role of forestry assumed more importance resulting in misuse of forest resources. Some of the contributing factors include:

- The present high population growth rate
- The lack of understanding of the characteristics
- Structure and dynamics of forest resources
- The scarcity of arable land
- The outlook of the people towards forests

The problem of forest management is tied with the conflicts between policy objectives developed by different government institutions for the utilization of land carrying important resources. Hence most scholars tend to argue for a comprehensive land use management plan while attempting to handle the enormous variety of ecological, social and environmental problems. However, though most governments are embarking on extensive forest conservation programs, there is a need for the people to be well informed on the role of forests in national development and alleviating the impacts of global warming and climate change. The lack of awareness on the dangers of forest depletion makes the people belittle the role of forests (García et al, 2017).

All efforts should be focused on rehabilitating formerly forested areas, both communal and private land. In general, most studies have clearly identified the problems of forest depletion both locally and internationally. There is a common agreement that forest resources depletion globally is a threat to the wellbeing and often the survival of a large part of the world's population. The principal question that needs to be asked throughout is how to conserve and where possible renew the forest resources in a sustainable manner. Afforestation has been a crucial aspect of most studies, mainly directed on the benefits of forests, especially in alleviating the continuous impacts of global warming and climate change (García et al, 2017).

Some of the policy recommendations that should be put forward include:

- An increase in public awareness and participation in the rehabilitation of more forest areas
- An increase in investments in the forest sector

Most proposals put forward tend to be directed towards the production of forests, hence leaving behind maintenance and management processes. Under such circumstances, forest depletion will automatically continue. In an attempt to control forest depletion, all three processes should be tailored together. These processes include:

- Forest production
- Forest utilization
- Forest maintenance and management

This is currently lacking both locally and internationally. Therefore, the government should formulate a framework which will integrate these processes together. This framework should act as a guiding principle in production, utilization, and management of forest products.

## 9.9 REFORMING THE INTERNATIONAL INSTITUTIONS THAT DEAL WITH IMPACTS OF GLOBAL WARMING AND CLIMATE CHANGE

International institutions, such as the United Nations Framework Convention on Climate Change (UNFCCC) and Inter-Governmental Panel on Climate Change (IPCC), and World Bank are critical to the mitigating and alleviating the impacts of global warming and climate change in the world. The effectiveness of these institutions within the present global context is, however, a major topic for debate. The 21$^{st}$ century offers different challenges compared to the last 60 decades. This is a major problem for these international institutions because their structure remains the same despite the changing global demands. The failure to adapt to present global challenges creates a need for reform of international institutions to reverse their decline. Reforming their basic structures offers the best solution to their current problems, and is a giant step to restoring their credibility in tackling various activities mandated to them (IPCC, 2014).

The importance of international institutions, such as the UNFCCC, IPCC, and World Bank in mitigating the impacts of global warming is evident in the success achieved since their creation. However, an increase in problems facing global institutions in recent years is becoming a major source of concern. There are many debates questioning their legitimacy as the UNFCCC, IPCC, and World Bank struggle to address existing and new global challenges in climate changes. A lack of extensive structural reforms since their creation is the root cause of the problem (Commonwealth Secretariat, 2009).

The global context has changed significantly since the creation of international institutions. They became operational following the Second World War and their structures focused on alleviating various impacts of global warming such as increased temperatures, erratic rainfall, increased relative humidity, drought and famine. The biggest problem is that global circumstances have changed but the institutions retain their basic structure (Nagel, 2005). Using the same model to tackle new threats creates new problems, which is currently the case.

Reforming institutions, such as the UNFCCC, IPCC, and World Bank requires conclusive and valid reasons to support them. One of the biggest criticisms is that global institutions are creating a modern form of colonialism by giving financial assistance with drastic repayment conditions in their Structural Adjustment Policies (SAPs). They play a prominent role in creating inequality between the rich and poor countries in the effect to fight climate change. Reforming these institutions is necessary because poor countries need access to financial support to fight the impact of global warming without the baggage of limiting conditions. Imposing conditions is not the mandate of international institutions. It indicates a flawed structural model that needs reforming (Commonwealth Secretariat, 2009).

Creating accountability should be an integral part of reforming international institutions, such as the UNFCCC, IPCC, and World Bank. The existing structural models lack accountability with the institutions opposing independent evaluation of their policies in the past. Monopolizing the operations by alienating stakeholders creates negative perceptions of the institutions. Reforms are therefore necessary for international institutions to ensure they meet democratic standards. Overall, the case for reforms has enough merit based on the shortcomings of global institutions, such and the UNFCCC, IPCC, and World Bank (Milner, 2005).

The most challenging aspect of reforming international institutions is determination of the areas to reform. The solution is analyzing and understanding the present global challenges. Among the most prominent is the emergence of new economic powers. China and India are emerging as global economic powerhouses after steady growth in recent decades. Population growth is also changing the demographic balance with most of it occurring in the developing nations. It is imperative that international institutions adopt and implement reforms to deal impacts of global warming and climate change based on global demographic and economic changes (IPCC, 2014).

The process of globalization is another factor to consider when determining the type of reforms to implement at international institutions. Institutions, such as the UNFCCC, IPCC, and World Bank have to operate in a wider and more complex field unlike in the past. They have to deal with emerging issues of concern, such as environmental conservation, water, and food security. It is a huge task to manage the different aspects of globalization without reforming the structures of international organizations with regard to enhancing cooperation (Commonwealth Secretariat, 2009).

The fact that global organizations are ineffective as individuals and as groups are proof of the need to develop and implement changes that facilitate coordination and competence.

The challenges of the 21st century require a collective effort to overcome as evident in the increasing number of summits held to address various issues. International institutions are failing to overcome issues, such as, hunger and poverty attributed to impacts of global warming because they have ineffective structures. Existing structures served their role in the past but the interdependence of present issues requires a different course of action. The relationships that exist between global issues make it impossible to separate them. Having concluded that there is a need to reform international institutions, the next step is determining the type of reforms. Below is a detailed analysis of proposed institutional reforms to improve the effectiveness of global organizations (Commonwealth Secretariat, 2009).

There are several ways to reform international institutions to alleviate the current and future impacts of global warming and climate change.

They include:

- Creating new institutions
- Reforming existing organizations
- Harnessing the collective effort of institutions
- Improving leadership to facilitate coordination among institutions

The specific mandates of the reforms include:

- Creating representation
- Enhancing the efficiency of global institutions
- Providing a framework for coordination

Restoring the legitimacy of international organizations requires them to represent the global state. It involves allocating responsibilities to countries based on their global influence. The shift in economic power over the years needs to be a factor for consideration when allocating responsibility to countries in dealing with the impacts of global warming. This means giving powerhouses bigger roles while protecting the interests of minor countries. The problem with the current setup is that minor countries have little to no input in global governance. Effective representation would help restore the credibility of international institutions, lost through a democratic deficit. Decision-making should no longer be the preserve of international bureaucrats but should include minor stakeholders, such as the civil society (David & Howden, 2013).

The second reform agenda is to make international institutions, such as the UNFCCC, IPCC, and World Bank effective. Their work needs transparency and accountability. This requires global organizations to fulfill their mandate according to existing and emerging global needs. Increasing the effectiveness and accountability of global organizations will play a major role in legitimizing their existence (Watts, 2013).

The third goal of reforming international organizations is to facilitate global cooperation in tackling global issues. The existing network of international organizations faces criticism for its neglect of developing countries. According to Watts (2013), the current setting has an imperial character with organizations working together to realize the interests of powerful nations. Collective action needs to be consistent and not undermine the autonomy of countries. Financial institutions, such as the World Bank need to reform their conditionality policies to create a platform for national and international leaders to work together to solve economic, social, and environmental issues. Reforming global organizations is a gradual and challenging process. The measures discussed below may play a big role in restoring the effectiveness and legitimacy of international organizations.

Creating new institutions is the least desirable measure. It involves eliminating institutions that no longer contribute to the global cause and creating organizations that meet the specific needs of the 21st century. The disadvantages of eliminating existing institutions far outweigh its benefits. This is with regard to opposition from countries with political and economic interests in them. The best solution is to form specific organizations for issue-areas similar to the United Nations programs. Creating new organizations from scratch is a complex and lengthy process. The solution to the current problem lies in creating subsidiaries of existing international institutions to deal with specific global demands.

Reforming international institutions requires them to undergo structural changes. This is necessary for most considering their failure to address issues of concern. They should play a passive role in reforming the institutions of recipient countries for example by giving advice (Watts, 2013).

Restoring democracy in the global state offers the best solution to combat the perception of imperialism within international institutions. Changing the policies and processes that govern them would help reverse the negative perception that international organizations serve specific interests. It involves establishing processes that include minority states in the negotiation and decision- making processes (Commonwealth Secretariat, 2009). Reducing bureaucracy in international institutions would also help them fulfill their mandate more effectively. It would enable development banks to revert to their role in facilitating development. The World Bank and other financial institutions should stop prioritizing lending and focus on economic and social development. They can achieve this objective by eliminating the overlap between the World Bank and regional developmental banks. Such reforms would enable international institutions to have a greater impact in marginalized regions, such as third world countries (Commonwealth Secretariat, 2009).

The best way to implement reforms in international institutions is through mobilization and engagement. The present and future global demands necessitate the need for alliances and cooperation among institutions. Leaders of international organizations need to reach out in a transnational network of global institutions. They have to create a platform for consultation and coordination to harness the capabilities of individual institutions and stakeholders. Important players in global development, such as national governments and heads of states also need to support international institutions in the decision-making processes. They have a better understanding of regional programs and this knowledge would enable international institutions to function more effectively (Watts, 2013).

Overall, international institutions, such as the UNFCCC, IPCC, and World Bank are underperforming based on various reasons. This affects their legitimacy negatively and is the reason why reforms are necessary. Reforms are costly and take time to achieve their desired outcome. It is crucial that international organizations reform their structures to enable them to deal with existing and future global issues effectively, from individual and collective perspectives.

## 9.10 KEY CHALLENGES IN MITIGATING THE IMPACTS OF GLOBAL WARMING

Acceptance of the essential process of global warming has continuously grown and is increasingly spreading. Mitigation of global warming involves actions aimed at limiting the degree of long-term climate change. The actions involve the reduction of human emissions of greenhouses gasses. The limitation can be accomplished through expanding the limit of carbon sinks through reforestation and eliminating petroleum derivatives by changing to low-carbon vitality sources, for example, atomic or sustainable power source. Despite the ever-increasing rates of global warming and the projected catastrophic predictions of these effects, it is still difficult for nations to mitigate the challenges of global warming (White, 2012). The difficulties in mitigating global warming include:

### 9.10.1 Politics and Economies of Nations

Nations' politics and economics play a major role in why mitigating global warming is difficult mainly because it has proved difficult to get all 194 countries to consent to re-instrument the worldwide vitality framework that drives their economies. Without a worldwide arrangement aimed at formulating principles geared towards reducing global warming that targets all nations to pull in weights in ensuring that the policies are followed is a challenge. There is, therefore, the need for great leadership from developed countries. Nevertheless, with developed countries transfixed by the financial emergency, there is next to zero drive to handle the atmosphere emergency.

### 9.10.2 Low-carbon Energy and Fossil Fuels

Energy reduction is the most efficient ways of cutting greenhouse gas emissions. It is proven that the use of renewable energy such as solar or wind energy is the cheapest ways to bring electricity to areas that lack the amenity (White, 2012). However, the use of energy-saving fuels is least embraced. This is mainly because despite renewable energy being unlimited, the technology appears costlier as compared to its alternatives. Moreover, in the year 2010, the fossil fuel industry received $409 billion in hand-outs as compared to the $66 billion received to promote clean energy. There is, therefore, a need to reverse such an imbalance.

### 9.10.3 Significant Time Lag in the Climate System

It takes a very long period of time in terms of decades or millenniums for a few parts of the atmosphere to react completely to changes in the atmospheric greenhouse gasses concentration. This is on the grounds that the world's seas can store a lot of warmth thus making it longer for the climatic system to warm up and respond to the concentration of the gasses in the atmosphere (Watts, 2013). In this manner, an inability to diminish ozone-depleting substance discharges sooner rather than later will secure a specific measure of future environmental change for a considerable length of time, if not a very long time to come. Due to this lag in time-frame, the actions taken to counteract the climate change are not as immediate as need be.

Global warming will affect every aspect of society. For the response to global warming through mitigation and adaptation to be effective, it has to involve every individual within the society. Despite the difficulties in global warming mitigation, such as politics and economies of nations, low-carbon energy and fossil fuels, significant time lags in the climate system. There is a need for taking stern actions to reduce the effects of climate change. Moreover, through adapting to the changes, the present generation will improve its current condition and accrue benefits such as quality life and public health, while helping future generations in forecasting and planning.

## Final note:

This chapter discussed challenge and approaches to overcome them to lower the phenomenon of global warming.

# Global Environmental Policy and Management To Deal With Impacts of Global Warming

# 10

CONTENTS

10.1 Introduction........................................................................210
10.2 International Environmental Agreements ......................................210
10.3 Economy, Environment and The Social Life of Individuals.............213
10.4 Environment Impact Assessment (EIA)..........................................213
10.5 Steps In Environmental Impact Assessment ...................................213
10.6 Policy Evaluation ........................................................................214
10.7 Advantages of the Global Environment Management Agreements........................................................................217
10.8 Challenges That Nations are Facing When Trying to Comply With The International Agreements ................................218
10.9 The Roles of International Financial Organizations in Curbing Environmental Challenges ..............................................219
10.10 Ethical Decision Making on Environmental Issues.......................220

## 10.1 INTRODUCTION

Environmental law is the network of regulations, statutes, customary and other common laws that address the impacts of various human activities on the environment. On the other hand, environmental policy refers to necessary steps, rules and regulations that are meant to govern the relationship that exists between human beings and the natural environment. Environment basically means anything that surrounds human beings. Most of the natural resources surrounding human beings are essential for survival and they include:

- Water sources
- Mountains
- Forests
- Animals
- Atmosphere

The benefits that humans get from the environment provided resources calls for the implementation of appropriate global environmental policies with the main aim of protecting the environment from adverse damages which would lead to negative effects than many would not withstand. This relationship is meant to benefit both human beings and the environment which helps to reduce the challenges that might arise from poor environmental practices by human beings (Charles and March, 2011).

## 10.2 INTERNATIONAL ENVIRONMENTAL AGREEMENTS

Different countries of the world have united to come up with policies that are meant to protect the environment from any form of destruction that would lead to the depletion of many natural resources that are meant to benefit human beings. These environmental management policies are also based on the forecast that showed that natural resources are gradually depleting with time and countries need to come up with measures to curb the challenges resulting from environmental degradation. This called for the United Nations and other environmental agencies to come up with strategies to protect and manage the environment (Charles and March, 2011).

The international environment agreements are based the analysis of the interrelationships that exist between human beings, plants, and other animals. Most of the environmental policies are meant to clean up the soil, waters, sediments, forests and the general environment surrounding the human habitation from pollution and other contaminants.

There are more than 2,000 international environmental management instruments that have been identified by Canada research on international political economy. These instruments have been represented in many conventions and most of them have been accepted by many countries to help in protecting the environment from degradation.

**Table 10.1:** Instruments for managing and protecting the environment

| Instruments | Characteristics |
|---|---|
| The convention on Nuclear safety | • This convention was held in Vienna in 1994<br>• The main aim was to protect the atmosphere from harmful substances which most of the factories in many countries emit in an uncontrolled way.<br>• It addressed issues such as EMEP protocol, heavy metals protocol, Gothenburg protocol, nitrogen oxide protocol, Sulfur emissions protocol and volatile organic compounds protocol.<br>• These harmful emissions are reported to cause diseases to human and wild animals hence the need to protect them.<br>• The argument was on the need to have a beneficial environment for both humans and other living organisms. |
| Geneva protocol | • Geneva protocol was basically meant to disallow the use of poisonous gases and other bacteriological methods of warfare during wars.<br>• The use of poisonous gases had adverse effects on the environment and this called for the need to have a protocol to be used in wars. |
| United Nations Convention on combating desertification | • This convention targeted the countries affected by drought and famine because of desertification.<br>• Africa was reported to be most affected and various measures needed to be implemented to mitigate these effects and of the measures advocated for included action programs by countries which were meant to combat desertification in those countries.<br>• The convention resulted in an agreement that became effective in 1996 and it consisted of the European Union and 195 states.<br>• Nevertheless, some countries such as Canada withdrew from the convention but later recalled it after realizing the advantage of being in such a convention. |

| | |
|---|---|
| North American Agreement on Environmental Cooperation | • This agreement was made between Mexico, Canada and the United States of America<br>• The agreement was in effect as from January 1994 and it was implemented because of the adverse effects that environmental degradation was reported to cause in the three countries.<br>• It consists of rules and regulations governing the protection and conservation of the environment. |
| European agreement on international carriage of harmful goods by inland waterways. | • This agreement was made in 2000<br>• It consisted of several European countries.<br>• It was meant to protect the waterways that are habitats of many animals which were reported to be endangered because of harmful goods that were transported through these inland waterways. |
| Convention on wildlife preservation and nature prevention in Western Hemisphere | • The convention was held in Washington DC and it was effective from 2004.<br>• This convention was meant to protect Americans who needed to preserve and protect their habitats.<br>• Flora and Fauna were reported to be endangered with time hence there was a need to protect them for the benefit that they had to Americans.<br>• Some of the issues addressed in the convention were the protection of national parks, national reserves, nature monuments, migratory birds among others. |
| International Treaty on plant genetic resources for agriculture and food | • This treaty was signed in 2001 in Madrid and it was effective as from the year 2004.<br>• It is also referred to as the international seed treaty that is meant to guarantee food security through the exchange, sustainable use and conservation of genetic resources. |

**Source:** https://www.czp.cuni.cz/knihovna/undp/modra/M14_kovar2.htm

Most of these treaties and agreements made between countries on best policies to protect and manage the environment have been implemented and this implies that environmental protecting and management has been given priority in most countries. Pollution of the environment has been criticized by many countries and policies that have been put in place to prosecute all those against these policies. Environmental organizations such as UNEP have ensured that all the policies that have been agreed upon by countries are put in place and necessary resources have been provided to support the implementation of the environmental policies (Sheldon and Kraft, 2013).

## 10.3 ECONOMY, ENVIRONMENT AND THE SOCIAL LIFE OF INDIVIDUALS

The agreements, protocols, conventions and treaties made between countries on issues related to the environment have shown the relationships between the economy, environment and the social life of individuals.

## 10.4 ENVIRONMENT IMPACT ASSESSMENT (EIA)

To implement these agreements and treaties, the potential threats have been considered since some of these strategies adopted by countries to protect and manage the environment have side effects on the people. The environmental impact assessment (EIA) evaluates and predicts the environmental effect that a proposed project has and it gives the appropriate recommendations. Alternatives are also provided to reduce the negative impacts that these environmental protection programs have on the environment. EIA has defined actions and their activities are all procedural instead of being subjective.

## 10.5 STEPS IN ENVIRONMENTAL IMPACT ASSESSMENT

The environmental impact assessment usually follows the following steps;

- Screening of the project to identify the potential impacts that it could have on the environment.
- Organizing a study team to review the proposed project.
- Scoping to help in identification of breadth and depth of the project.
- Analyzing the impacts that need to be assessed, identifying the extent of the impacts of giving the recommendations.
- Giving alternatives to the proposed project if the impacts assessed are beyond control.
- Planning for the mitigation measures to reduce, prevent and compensate for these impacts.
- Action planning to determine how feasible the alternatives are to the project.
- Engaging the public to give their views and opinion concerning

the project and the effects that they might have on the environment if implemented.

- Making decision pertaining to how the project will be implemented and the necessary steps in mitigating the effects if any.
- Follow up to identify if the set goals and objectives are met.

## 10.6 POLICY EVALUATION

The global community through international organizations such as United Nations (UN) and United nations environmental programs have identified that the methods that most of the individuals, factories and other organizations are using to dispose their waste is the key factor contributing to environmental degradation. The heated argument is on the best method to dispose of waste products and the two main options were landfilling or burning. This issue affected most of the treaties and agreements since some countries had different opinions. However, most of them agreed on the burning of waste products under given conditions to help in managing and protecting the environment (David and Howden, 2013).

### 10.6.1 Garbage Disposal

The treaties, agreements, protocols and conventions addressed issues related to garbage disposal methods, including burning and landfilling, which have been reported to have several pros and cons and this calls for the identification of the best waste management technique. All humanly possible efforts have been applied to ensure that the waste management technique that organizations and individuals adopt, have minimum negative effects on the environment (David and Howden, 2013).

Given two options, burning or landfilling waste, the best method as per the agreements made between countries in UN is burning. The method is always criticized by many countries, people and organizations but it is effective depending on the method adopted in burning. Furthermore, not all waste that is burnt has negative effects on the environment.

One of the advantages of burning waste, as reported by United Nations environmental program (UNEP), is that it creates more space to the environment which could be used for productive activities. The huge heaps of waste in many cities and other waste disposal areas have consumed a lot of space that can be used for development projects. Burning clears the area previously occupied by garbage and this creates a conducive environment which is the aim of all environmental management agencies (Rao et al, 2016).

Incineration should be applied in burning these wastes to ensure that all the gaseous emissions to the atmosphere are treated to avoid harmful effects to human beings, plants and other living organisms. Burning, by incinerating involves treating the gaseous emissions before releasing to the atmosphere and it is preferred to the land filling which leaves the environment unpleasant and wastes space that could be used for other productive activities (Sheldon and Kraft, 2013).

Countries have argued that incineration plants that are used to burn waste products generate energy. This energy can be used to generate heat and electricity which are useful in many homes and organizations. This saves on expenses that are incurred in importing electricity and heat from other countries. The demand for electricity and heat has been reported to be very high such that the electricity producing companies don't meet the demand in many developing countries. The incineration process of the garbage generates heat and electricity that partially helps in meeting demand.

The burning of waste products is thus preferred to landfilling in that it generates energy that can be used at homes for lighting, operating electrical appliances and heat products used for cooking. This gives a sufficient reason to support the burning of garbage instead of landfilling.

Environmental agencies have reported that garbage disposal areas are usually located far from the organizations and this implies that transport costs of this garbage must be incurred which reduces the profits of the organizations. To minimize these costs, the organizations need to find alternatives to dispose of their waste. Burning could be a solution to this challenge since the organizations will no longer need to transport this waste which will in return save their money (Sheldon and Kraft, 2013). The costs of transporting garbage could be used in the development of a city, a country or the organization itself. This gives us a reason to do away with landfilling technique of garbage disposal.

Treaties and agreements made between countries are also concerned with the underground water in that it should not be contaminated in any way since it acts as a source of water for many individuals and organizations.

Landfilling, as a garbage management technique, contaminates this water in that the waste on the earth's surface seeps through the soil and the underground water is contaminated (White, 2012). Consumption of this water is harmful to human beings, plants and other living organisms. The burning of waste products ensures that the underground waters are safe for consumption and this makes the method efficient in disposing of garbage since it favors our waters.

Rivers, lakes and oceans are also affected by garbage filled in people's lands in that the rains might wash the waste and take it to these water sources. The water sources are hence unfit for human consumptions and it might also kill the living creatures in these water sources. The adoption of the burning of garbage, as a waste management technique, is thus preferred to land filling technique (White, 2012).

For a better waste management technique, the burning of wastes by incineration makes it easier to manage lots of wastes that individuals and organizations produce. It has been reported that 90% of this waste products can be disposed of by burning which implies that the method is efficient enough compared to landfilling. The land filling strategy only helps in organic decomposition which might not be efficient since not all wastes are biodegradable. This leaves individuals and organizations, who need to dispose of their waste, with only one option of burning their waste. This ensures that they dispose of their wastes fast and it also conserves their environment.

United Nations (UN) and the United Nations Environmental Program (UNEP) are interested in conserving, protecting and managing the environment. They have agreed that waste management and disposal techniques should be chosen in a way that the effect the method has on the environment is considered (Charles & March, 2011). The method chosen should also be cost effective and should be easy enough to help in disposing garbage within the shorted time possible. Burning of garbage is thus preferred according to many environmental conventions held since it is observed to take care of our environment especially when done by incineration. However, some countries have opted to disagree with this decision and they have decided to settle on their own method.

## 10.6.2 Cultural Properties that have been endangered

A close analysis of the environment showed that the cultural properties have been endangered to different extents in different continents and this gives a clear picture of how policies need to be implemented in different continents to protect many indigenous species.

## 10.7 ADVANTAGES OF THE GLOBAL ENVIRONMENT MANAGEMENT AGREEMENTS

There are numerous benefits that accrue to countries which agree to mutually protect and manage the environment.

- When countries in a region join to come up with policies to help in protecting and managing the environment, they learn on ways to make the environment better for habitation.
- Member countries are equipped with the best strategies and measures to take care of the environment and the best ways to deal with people who do not comply with the environmental policies. This is an advantage over other countries that just depend on their own on matters related to the environment.
- Global environmental management agreements are a source of financial support to countries in with challenges of implementing environmental policies. Member countries benefit from grants to support all environmental management programs which many could not manage on their own. Developing countries benefit most and this helps to protect their natural resources that are used in most of the productive activities.
- All countries are concerned with protecting their environment and this implies that they will also ensure that they have strict rules to protect the countries from any other party, from other countries, that tries to make their environment unpleasant. The global organizations such as UNEP impose strict rules and penalties to countries which are found degrading the environment which affects even other countries.
- Some countries have been reported to emit poisonous gases to the atmosphere which extend to other countries which endangers the lives of residents in more than one country. The global organizations help to curb this issue by implementing policies that ensure that all activities that a country engage in do not in any way affect the environment of neighboring countries.
- Education support on matters related to environmental management and protection is another benefit that accrues to countries in global organizations that support environmental protection and management. The organizations offer scholarships to most of the countries where eligible candidates are educated all

matters related to measures to protect the environment, mitigating the impacts of the environmental management policies and the importance of conserving and protecting the environment. When these people are educated on such important matters related to the environment, they help the country to educate the community members on the ways to protect the environment and the importance of protecting it. This is an advantage to countries that have joined global organizations that are concerned with protecting and management of the environment.

## 10.8 CHALLENGES THAT NATIONS ARE FACING WHEN TRYING TO COMPLY WITH THE INTERNATIONAL AGREEMENTS

Despite the vast benefits associated with complying with the international agreements on environmental protection and management, some countries are still unable to be part of the international environmental management programs. Some of these challenges result from their arguments about the need for environmental management among other reasons.

- One of the challenges facing many countries is the lack of resources to comply with the proposed environmental management policies. Most of the strategies require a lot of resources that most of the developing countries cannot afford. This makes them fail to comply with the internationally agreed environmental policies even though these countries really need to protect their environment.
- Some of these countries start implementing the policies provided by the global environmental programs and some ends up failing in the middle of the process which makes the policies ineffective.
- There are some countries that completely differ from some policies that countries propose. They tend to assume that the policies have negative effects on their environment which gives them sufficient reason to differ with these agreements.
- Some countries also tend to dominate in everything such that whenever their proposals on environmental management and protection policies are rejected, they quit from the global unions. This has become a challenge and countries have faced several challenges resulting in environmental degradation due to ignorance and illusion of knowledge.

## 10.9 THE ROLES OF INTERNATIONAL FINANCIAL ORGANIZATIONS IN CURBING ENVIRONMENTAL CHALLENGES

International financial organizations such as the World Bank have been at the forefront in helping many developing nations to solve some of the challenges that they are facing. These international financial organizations help by offering loans and Grants to governments of the affected nations to help them implement projects that will solve challenges that they are facing. Some of the projects that an organization like World Bank supports include:

- Construction of infrastructure such as dams, water tunnels, boreholes
- Afforestation
- Education
- Other community development programs

One of the benefits that an organization like the World Bank would get from helping developing nations is that it saves on costs that would be incurred because of the adverse effects of environmental degradation. The projects that this organization supports are basically meant to help in protecting and managing the environment. They save by protecting the environment from degradation which is so costly to restore. Arid and Semi-arid areas benefit from boreholes which can be used in supporting agricultural activities which help residents of those areas in sustaining their lives. If no help was offered by the World Bank, a lot of money would be used in relief foods and this helps the organization to save on such expenses (Sheldon and Kraft, 2013).

World Bank also helps in environmental protection and management by facilitating research related to the environment which helps to forecast the extent to which the environment has been degraded with the aim of finding measures to mitigate any negative effects. The World Bank is always concerned with the environment which affects the economies of many countries because of poor environmental management practices. The countries which have destroyed their environment by means such as deforestation, emitting harmful to the atmosphere and poor garbage disposal techniques will always need help to raise their economy and this becomes a burden to the World Bank. Facilitating research on our environment helps to identify ways through which the environment can be protected and managed for the benefit of human beings and other living creatures (Sheldon and Kraft, 2013).

## 10.10 ETHICAL DECISION MAKING ON ENVIRONMENTAL ISSUES

Despite the fact that the environmental regulations are lagging behind scientific evidence on this particular evidence does not, therefore, purport massive wastage of toxic materials into the environment. Therefore, under this environmental perspective, people should do an analysis of whether industries should follow an economic approach usually based on utilitarianism perspective. It should be understood that human life value can never be reduced solely to be equated to economic value. Besides, non-human nature only serves as instrumental values justifying the wellbeing of human life, and for that reason, traditional environmental ethics should serve the function of protecting environmental lives regardless of human activities.

### Final note:

This chapter discussed various approaches to handle the challenge of global warming along with roadblocks and their potential solutions. "Find your place on the planet. Dig in, and take responsibility from there." — Gary Snyder; in lieu of this quote each citizen of this planet should do his/her role to push for reforms and strategies aimed at tackling the challenge head-on and hand over clean earth to the coming generations.

# REFERENCES


1. Anderson, M. (2015). Global Warming. New York: The Rosen Publishing Group, Inc, .
2. Andrews, A. (2012). Nuclear Power Plant Design and Seisminc Safety Consideration. Darby, Pennsylvania: Diane Publishing.
3. Ariel, D. and Robert, O. (2011). Handbook on Climate Change and Agriculture. London: Edward Elgar Publishing.
4. By Gregory Bernard,C., Egnin, M., and Bonsi, C. (2017). Nematology - Concepts, Diagnosis and Control. Intech Open. Edited by Mohammad Manjur Shah and Mohammad Mahamood.
5. Bhupinder, P.S, Annette L.C, & Chan. (2011). Soil Health and Climate Change. Berlin: Springer Science & Business Media.
6. Casper, J. K. (2010). Changing Ecosystems: Effects of Global Warming. New York: Infobase Publishing.
7. Novak,C.A, Van Giesen,E., and DeBusk, K.M. (2014). Designing Rainwater Harvesting Systems: Integrating Rainwater Into Building Systems. New York: John Wiley & Sons.
8. . Eccleston,C.H.and March, F. (2011). Global Environmental Policy: Concepts, Principles, and Practice. New York: CRC Press.
9. Goldman,C.R., Kumagai,M., and Robarts, R.D. (2012). Climatic Change and Global Warming of Inland Waters: Impacts and Mitigation for Ecosystems and Societies. New Jersey: John Wiley & Sons.
10. Bjorkman,C. and Niemela, P. (2015). Climate Change and Insect Pests. London: CABI.

11. Bryant C.R, Sarr,M.A and Délusca, K. (2016). Agricultural Adaptation to Climate Change. Berlin: Springer.
12. Hoanh, C.T., Smakhtin,V, and Johnston, R (2015). Climate Change and Agricultural Water Management in Developing Countries. London: CABI.
13. Commonwealth Secretariat. (2009). Reform of International Institutions: Towards a Commonwealth Agenda: Issue 85 of Economic paper, Commonwealth Secretariat. London: Commonwealth Secretariat.
14. Daniel, H. & Cynthia, R. (2011). Handbook of Climate Change and Agroecosystems: Impacts, Adaptation, and Mitigation. Singapore: World Scientific.
15. White, D.H. and Howden, S.M (2013). Climate Change: Significance for Agriculture and Forestry: Systems Approaches Arising from an IPCC Meeting. Berlin: Springer Science & Business Media.
16. Dinar, A., Hassan,R., Benhin,J. and Mendelsohn, R. (2012). Climate Change and Agriculture in Africa: Impact Assessment and Adaptation Strategies. London: Earthscan.
17. Dinda and Soumyananda. (2015). Handbook of Research on Climate Change Impact on Health and Environmental Sustainability. New York: IGI Global.
18. Karlen, D.L. and Rice, C.W. (2018). Enhancing Soil Health to Mitigate Soil Degradation. Basel: MDPI.
19. Eastwood, E. (2011). Global Warming: What Else Can You Do About It? Indiana: iUniverse.
20. Wall, E. B. S. (2011). Farming in a Changing Climate: Agricultural Adaptation in Canada. Canada: UBC Press.
21. FAO. (2013). Tackling Climate Change Through Livestock: A Global Assessment of Emissions and Mitigation Opportunities. London: Food & Agriculture Org.
22. FAO. (2013). The State of the World's Land and Water Resources for Food and Agriculture: Managing Systems at Risk. London: Routledge.
23. Chevesich,G., , Neary, P., Daniel G., Scott, David F., Benyon, Richard G., Reyna, Teresa. (2017). Forest management and the impact on water resources: a review of 13 countries. London: UNESCO Publishing.
24. German Advisory Council On Global Change. (2010). New Structures for Global Environmental Policy. Berlin: Earthscan.

25. Hakeem, K. R. (2015). Crop Production and Global Environmental Issues. Berlin: Springer.

26. Haldar, I. (2010). Global Warming: The Causes and Consequences. London: Mind Melodies.

27. Haruna,I.M, and Usman, A. (2013). Agronomic Efficiency of Cowpea Varieties (Vigna unguiculata L. Walp) under Varying Phosphorus Rates in Lafia, Nasarawa State, Nigeria. Asian Journal of Crop Science, 5: 209-215.

28. Hennessy, T. (2017) The potential food safety, economic and environmental impacts of climate change on the dairy production chain on the island of Ireland, ISBN:Publication .

29. IAEA. (2013, June 29). Concluding Statement by the President of the Conference "Nuclear Power in the 21st Century". Retrieved Sep. 15, 2013, from International Atomic energy Agency: http://nuclearforum2013.org/media/news/1295/

30. Chabay,I., Frick,M., Helgeson, J (2015). Land Restoration: Reclaiming Landscapes for a Sustainable Future. New York: Academic Press.

31. IPCC. (2014). Climate Change 2014 – Impacts, Adaptation and Vulnerability: Global and Sectoral Aspects. London: Cambridge University Press.

32. Jha,S., Bacon,C.M, Philpott, S.M, Méndez, E.S., et al (2014) Shade Coffee: Update on a Disappearing Refuge for Biodiversity, BioScience, Volume 64, Issue 5, May 2014, Pages 416–428

33. Fuhrer, J and Gregory, P.J. (2014). Climate Change Impact and Adaptation in Agricultural Systems: Soil Ecosystem Management in Sustainable Agriculture. London: CABI.

34. Maharjan, K.L and Joshi, N.P (2013). Climate Change, Agriculture and Rural Livelihoods in Developing Countries. Berlin: Springer Science & Business Media.

35. Kumar D, Kalita P. Reducing Postharvest Losses during Storage of Grain Crops to Strengthen Food Security in Developing Countries. Foods. 2017;6(1):8. Published 2017 Jan 15. Lichtfouse, E. (2009). Climate Change, Intercropping, Pest Control and Beneficial Microorganisms. Berlin: Springer Science & Business Media.

36. Lord, N. (2011). Early Warming: Crisis and Response in the Climate-changed North. New York: Counterpoint Press.

37. Kang, M.S. and Banga, S.S (2013). Combating Climate Change: An

Agricultural Perspective. Florida: CRC Press.

38. Coll,M. and Wajnberg,E. (2017). Environmental Pest Management: Challenges for Agronomists, Ecologists, Economists and Policymakers. New York: John Wiley & Sons.
39. Tuteja,N. Gill, S.S and Tuteja, R. (2012). Improving Crop Productivity in Sustainable Agriculture. New York: John Wiley & Sons.
40. OECD. (2012). Farmer Behaviour, Agricultural Management and Climate Change. London: OECD Publishing.
41. Mahul, O and Stutley, C.J. (2010). Government Support to Agricultural Insurance: Challenges and Options for Developing Countries. London: World Bank Publications.
42. Ormandy EH, Dale J, Griffin G. (2010). Genetic engineering of animals: ethical issues, including welfare concerns. Can Vet J. 2011;52(5):544–550.
43. Otto, C. Doering III, Randolph, J.C, Southworth,J. and Pfeifer,R.A. (2012). Effects of Climate Change and Variability on Agricultural Production Systems. Berlin: Springer Science & Business Media.
44. Ouda, S. (2015). Major Crops and Water Scarcity in Egypt: Irrigation Water Management under Changing Climate. Berlin: Springer.
45. Zdruli,P., Pagliai, M., Kapur,S., Cano, A.F . (2010). Land Degradation and Desertification: Assessment, Mitigation and Remediation. Berlin: Springer Science & Business Media.
46. Pearson, C. S. (2011). Economics and the Challenge of Global Warming. London: Cambridge University Press.
47. Malik,P.K Raghavendra. B, Ti,J., Richard. K, Cadaba. S . (2015). Livestock Production and Climate Change. London: CABI.
48. Sengar, R.S. and Sengar, K. (2014). Climate Change Effect on Crop Productivity. Florida: CRC Press.
49. Rao, Prakash, Patil, and Yogesh. (2016). Reconsidering the Impact of Climate Change on Global Water Supply, Use, and Management. London: IGI Global.
50. Rattan Lal, B. R. (2014). Sustainable Intensification to Advance Food Security and Enhance Climate Resilience in Africa. Berlin: Springer.
51. Lal,R., Kraybill,D., Hansen,D.O., Singh, B., Mosogoya,T., Eik, L.O. (2016). Climate Change and Multi-Dimensional Sustainability in African Agriculture: Climate Change and Sustainability in Agriculture.

Berlin: Springer.

52. Reddy, P. P. (2014). Climate Resilient Agriculture for Ensuring Food Security. Berlin: Springer.
53. Reynolds, M. P. (2010). Climate Change and Crop Production. London: CABI.
54. Schrope, M. (2013, April 2). Nuclear power prevents more deaths than it causes. Retrieved Sept. 14, 2013, from Chemical and Engineering news: http://cen.acs.org/articles/91/web/2013/04/Nuclear-Power-Prevents-Deaths-Causes.html
55. Kamieniecki, S. and Kraft, M (2013). The Oxford Handbook of U.S. Environmental Policy. New York: OUP USA.
56. Shrader-Frechette, K. (2011). What Will Work: Fighting Climate Change with Renewable Energy, Not Nuclear Power. London: Oxford University Press.
57. Yadav,S.S, Redden,R., Hatfield,J.L, Lotze-Campen,H., Hall, A.J.W (2011). Crop Adaptation to Climate Change. New York: John Wiley & Sons.
58. Spellman, F. R. (2014). Environmental Impacts of Renewable Energy. Florida: CRC Press,.
59. Tutelyan, V. (2013). Genetically Modified Food Sources: Safety Assessment and Control. New York: Academic Press.
60. Veerasamy. S, John. G, Lance. B, Cadaba. P. (2015). Climate Change Impact on Livestock: Adaptation and Mitigation. Berlin: Springer.
61. Veerasamy. S, Naqvi S.K.M, Thaddeus. Ei, Jeffrey. L, Rattan. L. (2012). Environmental Stress and Amelioration in Livestock Production. Berlin: Springer Science & Business Media,.
62. Walter Leal Filho, Anthony O. Esilaba, Karuturi P.C. Rao, and Gummadi Sridhar. (2014). Adapting African Agriculture to Climate Change: Transforming Rural Livelihoods. Berlin: Springer.
63. Watts, R. G. (2013). Engineering Response to Climate Change, Second Edition. Florida: CRC Press.
64. Weaver, A. (2011). Generation Us: The Challenge of Global Warming. New York: Orca Book Publishers.
65. White, W. A. (2012). Biosequestration and Ecological Diversity: Mitigating and Adapting to Climate Change and Environmental

Degradation. Florida: CRC Press.

66. Winston, A. S. (2014). The Big Pivot: Radically Practical Strategies for a Hotter, Scarcer, and More Open World. London: Harvard Business Review Press.
67. World NuclearAssociation. (2013, August). Safety of Nuclear Power Reactors. Retrieved Sep. 14, 2013, from World Nuclear Association: http://www.world-nuclear.org/info/Safety-and-Security/Safety-of-Plants/Safety-of-Nuclear-Power-Reactors/
68. XinRong, Z and Ibrahim, D. (2016). Energy Solutions to Combat Global Warming. Berlin: Springer.

# INDEX


A

Adult population 144
Aggregate stability 86
Agricultural crop farming 41
Agricultural development 37
Agricultural education 63
Agricultural irrigation 42
Agricultural productivity 36
Agroforestry 69
Animal husbandry 37
Animal productivity 125
Antecedent soil moisture 192
Atlantic hurricane 3
Atmospheric temperature 150, 156

B

Bacterial contamination 13
Beef farming 118, 124, 126
Bimodal rainfall 189
Biodegradable fuels 59
Botanical composition 130

C

Calving rate 129
Camel production 147
Cane cultivation 102
Canopy development 108, 109
Carbon dioxide assimilation 78
Carbon monoxide 16
Cardinal temperatures 77
Cassava mosaic disease 149, 150, 151
Cattle husbandry 147
Chlorofluorocarbons (CFCs) 5
Chloroplast structure 45
Civil nuclear power 182
Climate change 2, 3, 4, 5, 10, 11, 12, 15, 16, 17, 24, 27, 28, 29, 30, 32
Climate system 3
Climatic constraint 37
Clump distribution 45
Coastal weather 2
Coffee berry disease (CBD) 152
Coffee production 155
Commercial energy 11, 12
Crop homogenization 64
Crop management 36, 39
Crop phenology 106
Cultivation 84, 85, 89
Cytoplasmic function 106, 111

## D


Dairy farming 118, 120, 123
Deep rooting nature 105
Deep rooting system 110
Deforestation 61, 62, 68
Dehydration 108
Depletion value 7
Desertification 63, 64, 65
Digestible energy (DE) 131
Discoloration 159
Dispersing activity 81
Diurnal variation 186
Domestic energy 181
Dry matter production 36
Dynamic equilibrium 83


## E


Economic control 63
Economic development 11
Economic feasibility 187
Electric power transmission 185
Endogenous rhythms 76
Energy management 178, 179, 180
Environmental deterioration 65, 67
Environmental impact assessment (EIA) 213
Environmental legislation 59
Environmental pollution 10
Environmental protection 56, 57
Environmental sustainability 57, 61
Enzyme ribulose bisphosphate carboxylase 45
Erosion control strategies 39
Erythrocytosis 121
ESS (Energy Storage Systems) 181
Evapotranspiration 36, 38, 47, 49, 50


## F


Fertilization 120, 140
Field capacity (FC) 192
Flood irrigation 172
Fodder soil conservation 41
Folic acid 155
Fragile ecosystem 64
Furrow irrigation 171, 172


## G


General Adaption Syndrome 121
Global warming 2, 5, 24, 25, 27, 28
Glucocorticoid hormones 121
Glycophytes 79, 82
Goat farming 118
Gradual reduction 12
Grain yield 97
Greenhouse effect 179, 181
Greenhouse gas 14, 17, 25, 26
Greenhouse methane gas 14


## H


Hastens maturity 109
Hatchability 149
Heat-resistant property 149
Hematological changes 121
Herbage digestibility 126
High ambient temperature 76
Human intervention 64
Human nutrition 122
Humidity 101, 104, 107, 112, 113
Hydraulic fracturing 13, 14, 15
Hydraulic fracturing technology 12
Hydrochlorofluorocarbons (HCFCs) 6


## I


Infiltration capacity 44

Infiltration process 44
Infiltration rate 173
Inter-Governmental Panel on Climate Change (IPCC) 201
Internally displaced (IDPs) 73
International Atomic Energy Agency (IAEA) 182
Inter-Tropical Convergence Zone (ITCZ) 39
In vitro organic matter digestibility (IVOMD) 131
Ion-resistant enzymes 82
Irrigation 103, 105, 111
Irrigation farming 173
Irrigation water 165, 166, 167, 168, 169, 170, 174

L

Larvae feeding 159
Leaf anthracnose 153
Leaf shedding 108
Livestock farming 118
Lower basal metabolism 120

M

Maintenance mechanism 15
Metabolic heat production 119, 120
Metabolism 80
Meteorological data 153
Milk production 118, 120, 122, 126, 139
Mitigating pollution 58
Modern consumerist society 56

N

National economic development 124
Natural carbon cycle 4
Natural gas production 12
Nematode development 160
Net-zero energy building (NZEB) 178
Nodulation process 109
Non-commercial energy 11
Nuclear energy 181, 182, 184
Nuclear engineering 185
Nutrient formation process 154
Nutrient ions 81
Nutrient management 88

O

Organic carbon 83, 84
Organic matter 82, 83, 84, 85, 86, 87, 90
Osmotic adjustment 108, 109
Oxidation-reduction reaction 76
Ozone layer depletion 6, 7, 8

P

Pantothenic acid 155
Percolation 37, 47
Periodic deficiency 127
Phenotypic stability 193, 194
Photosynthesis 40, 45, 49, 50, 52, 53, 54
Photosynthetically active radiation 109
Photosynthetic enzymes 82
Photosynthetic pigments 81
Pigeon pea production 108
Plantation 61
Plant-environment system 103
Pollution reduction 10
Potato production 98, 100
Poultry farming 118, 135, 137

Poverty 60, 65, 66, 68
Proline accumulation 108
Protein synthesizing mechanism 78
Protozoan disease 143

R

Radiant heating 78
Rain compatibility 44
Rainfall distribution 38
Rainfall intensity 43
Rainfall-yield relationship 97
Rapid tuber enlargement 99
Renewable energy 11
Reproductive efficiency 127, 129

S

Saline water 79, 82
Salt-resistant enzyme 82
Seed dormancy 98
Sexual maturity 143
Shale gas 12, 13
Social development 57
Soil erosion 129
Soil fertility 103
Soil moisture 38, 45, 46, 52
Soil salinity 79, 80, 81, 82, 85
Soil surface 14
Solar energy 58
Solar radiation 40, 46
Species distribution 46
Stability factor (SF) 193
Statistical analysis 43
Stomatal behavior 108
Stomatal conductance 78
Structural Adjustment Policies (SAPs) 202
Sustainable Europe Research Institute (SERI) 56

T

Tea production 100, 101
Thornthwaite system 46
Toxic gases 68
Translocation 108
Tropical climates 144
Tropical environment 76
Trypanosomiasis 148

U

Ultraviolet radiations 5
United Nations environmental program (UNEP) 214
Unpredictability nature 96
Urban Heat Island (UHI) 5
Urbanization growth 12

V

Vegetation cover systems 41
Volcanic ash 43

W

Wanton water evaporation 5
Waste management 61
Water balance 36, 46, 48, 49
Water harvesting systems 191
Water infiltration 44
Water vapor pressure 78
Whitefly population density 150